AF389397

Novel Strategies for the Development of Healthier Meat and Meat Products and Determination of Their Quality Characteristics

Novel Strategies for the Development of Healthier Meat and Meat Products and Determination of Their Quality Characteristics

Editors

Claudia Ruiz-Capillas
Ana Herrero Herranz

MDPI • Basel • Beijing • Wuhan • Barcelona • Belgrade • Manchester • Tokyo • Cluj • Tianjin

Editors
Claudia Ruiz-Capillas
Spanish National Research
Council, CSIC,
Department of Products
Spain

Ana Herrero Herranz
Spanish National Research
Council, CSIC,
Department of Products
Spain

Editorial Office
MDPI
St. Alban-Anlage 66
4052 Basel, Switzerland

This is a reprint of articles from the Special Issue published online in the open access journal *Foods* (ISSN 2304-8158) (available at: https://www.mdpi.com/journal/foods/special_issues/Novel_Strategies_Development_Healthier_Meat_Meat_Products_Determination_Quality_Characteristics).

For citation purposes, cite each article independently as indicated on the article page online and as indicated below:

LastName, A.A.; LastName, B.B.; LastName, C.C. Article Title. *Journal Name* **Year**, *Volume Number*, Page Range.

ISBN 978-3-0365-2984-4 (Hbk)
ISBN 978-3-0365-2985-1 (PDF)

Contents

About the Editors . . . . **vii**

Preface to "Novel Strategies for the Development of Healthier Meat and Meat Products and Determination of Their Quality Characteristics" . . . . **ix**

Claudia Ruiz-Capillas and Ana M. Herrero
Novel Strategies for the Development of Healthier Meat and Meat Products and Determination of Their Quality Characteristics
Reprinted from: *Foods* **2021**, *10*, 2578, doi:10.3390/foods10112578 . . . . **1**

Nieves Núñez-Sánchez, Carmen Avilés Ramírez, Francisco Peña Blanco, Pilar Gómez-Cortés, Miguel Ángel de la Fuente, Montserrat Vioque Amor, Alberto Horcada Ibáñez and Andrés Luis Martínez Marín
Effects of Algae Meal Supplementation in Feedlot Lambs with Competent Reticular Groove Reflex on Growth Performance, Carcass Traits and Meat Characteristics
Reprinted from: *Foods* **2021**, *10*, 857, doi:10.3390/foods10040857 . . . . **7**

Yong-Hong Feng, Song-Shan Zhang, Bao-Zhong Sun, Peng Xie, Kai-Xin Wen and Chen-Chen Xu
Changes in Physical Meat Traits, Protein Solubility, and the Microstructure of Different Beef Muscles during Post-Mortem Aging
Reprinted from: *Foods* **2020**, *9*, 806, doi:10.3390/foods9060806 . . . . **17**

Leo Nyikadzino Mahachi, Monlee Rudman, Elodie Arnaud, Voster Muchenje and Louwrens Christiaan Hoffman
Application of Fat-Tailed Sheep Tail and Backfat to Develop Novel Warthog Cabanossi with Distinct Sensory Attributes
Reprinted from: *Foods* **2020**, *9*, 1822, doi:10.3390/foods9121822 . . . . **27**

Magdalena I. Cerón-Guevara, Esmeralda Rangel-Vargas, José M. Lorenzo, Roberto Bermúdez, Mirian Pateiro, Jose A. Rodríguez, Irais Sánchez-Ortega and Eva M. Santos
Reduction of Salt and Fat in Frankfurter Sausages by Addition of *Agaricus bisporus* and *Pleurotus ostreatus* Flour
Reprinted from: *Foods* **2020**, *9*, 760, doi:10.3390/foods9060760 . . . . **45**

Dipak Kumar Banerjee, Arun K. Das, Rituparna Banerjee, Mirian Pateiro, Pramod Kumar Nanda, Yogesh P. Gadekar, Subhasish Biswas, David Julian McClements and Jose M. Lorenzo
Application of Enoki Mushroom (*Flammulina Velutipes*) Stem Wastes as Functional Ingredients in Goat Meat Nuggets
Reprinted from: *Foods* **2020**, *9*, 432, doi:10.3390/foods9040432 . . . . **61**

Pratap Madane, Arun K. Das, Mirian Pateiro, Pramod K. Nanda, Samiran Bandyopadhyay, Prasant Jagtap, Francisco J. Barba, Akshay Shewalkar, Banibrata Maity and Jose M. Lorenzo
Drumstick (*Moringa oleifera*) Flower as an Antioxidant Dietary Fibre in Chicken Meat Nuggets
Reprinted from: *Foods* **2019**, *8*, 307, doi:10.3390/foods8080307 . . . . **77**

Simona Grasso, Tatiana Pintado, Jara Pérez-Jiménez, Claudia Ruiz-Capillas and Ana Maria Herrero
Potential of a Sunflower Seed By-Product as Animal Fat Replacer in Healthier Frankfurters
Reprinted from: *Foods* **2020**, *9*, 445, doi:10.3390/foods9040445 . . . . **97**

Daniel Franco, Artur J. Martins, María López-Pedrouso, Laura Purriños, Miguel A. Cerqueira, António A. Vicente, Lorenzo M. Pastrana, Carlos Zapata and José M. Lorenzo
Strategy towards Replacing Pork Backfat with a Linseed Oleogel in Frankfurter Sausages and Its Evaluation on Physicochemical, Nutritional, and Sensory Characteristics
Reprinted from: *Foods* **2019**, *8*, 366, doi:10.3390/foods8090366 . . . . **111**

Tatiana Pintado, Claudia Ruiz-Capillas, Francisco Jiménez-Colmenero and Ana M. Herrero
Impact of Culinary Procedures on Nutritional and Technological Properties of Reduced-Fat Longanizas Formulated with Chia (*Salvia hispanica* L.) or Oat (*Avena sativa* L.) Emulsion Gel
Reprinted from: *Foods* **2020**, *9*, 1847, doi:10.3390/foods9121847 . . . . **123**

Leticia Aline Gonçalves, José M. Lorenzo and Marco Antonio Trindade
Fruit and Agro-Industrial Waste Extracts as Potential Antimicrobials in Meat Products: A Brief Review
Reprinted from: *Foods* **2021**, *10*, 1469, doi:10.3390/foods10071469 . . . . **135**

Claudia Ruiz-Capillas and Ana M. Herrero
Development of Meat Products with Healthier Lipid Content: Vibrational Spectroscopy
Reprinted from: *Foods* **2021**, *10*, 341, doi:10.3390/foods10020341 . . . . **145**

About the Editors

Claudia Ruiz-Capillas has a degree in Veterinary Medicine from UCM (Spain). She is currently a senior research scientist in the Institute of Food Science Technology and Nutrition (ICTAN-CSIC). She addresses various lines of research, including the development of healthier meat products (fat analogues, reduction of salt and additives, etc.); meat analogues; sensory characteristics and consumer studies; chemical, physical, and, structural changes in the components and properties of meat products; quality and safety characteristics with special reference to biogenic amines, nitrates, and nitrites; and the application of storage technologies (HP, protective atmospheres, etc.). She has participated in and led a number of national and international research projects. She has published over 155 peer-reviewed research articles, 18 book chapters, and numerous scientific outreach articles (h-index: 38). She has also been a guest co-editor of several Special Issues of journals and the co-editor of some books.

Ana Herrero Herranz has a degree in Chemistry from UCM (Spain). She is currently a scientific researcher at the Institute of Food Science, Technology, and Nutrition (ICTAN-CSIC). Her research focuses on the development of healthier meat products with reference to lipid content and certain bioactive compounds (minerals, fiber, etc.). This research uses various reformulation strategies to study the relationships between the physicochemical and textural properties of these products, as well as structural changes (using vibrational spectroscopy in situ). She has taken part in national and international projects and projects with private enterprises, acting as principal researcher. She has published articles in SCI journals (80) and scientific outreach journals (19) and has written several book chapters (15). She has also given courses and seminars, taught at the university level (degree and doctorate), and taken part in the direction of doctoral theses and training of students, technical personnel, etc.

Preface to "Novel Strategies for the Development of Healthier Meat and Meat Products and Determination of Their Quality Characteristics"

Meat and meat products are widely demanded and accepted by consumers. Health is one of the main criteria for consumers when choosing the type of food they eat. This, along with more sustainable global production, changes in lifestyles, and nutritional ideologies in pockets of the population, has led to the fact that today there is a growing demand in the market for healthier and more functional foods. All these factors are responsible for new trends and innovations in the development of new health products, among which are novel strategies for the development of healthier meats and meat products at different stages of the food chain from farm to fork. These strategies are based mainly on the reduction or elimination of unhealthy components, which may or may not be replaced by a healthy alternative. This is done while maintaining the same quality criteria demanded of traditional products in terms of their sensory, technological, nutritional, functional, safety, and other characteristics.

This Special Issue has collected nine original research articles, two reviews, and one editorial (https://www.mdpi.com/2304-8158/10/11/2578/htm) with all-new, relevant information. These papers cover different strategies for modifying meat and meat products to make them healthier and more sustainable and the different quality control methods for monitoring them.

This SI will prove very useful and interesting for all those involved in the meat and meat products industry from farm to fork (livestock producers, processing companies, researchers, scientists, consumers, administration, etc.). Lastly, many of the studies featured in this SI leave the door open to future research in this area of growing interest, not only for the meat and meat product industry but for other foods as well.

We would like to acknowledge the efforts of the authors of the publications in this Special Issue.

Claudia Ruiz-Capillas, Ana Herrero Herranz
Editors

Editorial

Novel Strategies for the Development of Healthier Meat and Meat Products and Determination of Their Quality Characteristics

Claudia Ruiz-Capillas * and Ana M. Herrero *

Institute of Food Science, Technology and Nutrition (ICTAN-CSIC), José Antonio Novais 10, 28040 Madrid, Spain
* Correspondence: claudia@ictan.csic.es (C.R.-C.); ana.herrero@ictan.csic.es (A.M.H.)

Citation: Ruiz-Capillas, C.; Herrero, A.M. Novel Strategies for the Development of Healthier Meat and Meat Products and Determination of Their Quality Characteristics. *Foods* **2021**, *10*, 2578. https://doi.org/10.3390/foods10112578

Received: 22 October 2021
Accepted: 24 October 2021
Published: 26 October 2021

Publisher's Note: MDPI stays neutral with regard to jurisdictional claims in published maps and institutional affiliations.

Meat and meat products are very popular foods and widely accepted by consumers. However, the global production and consumption dynamics of meat and meat products have evolved rapidly as the result of changing lifestyles and nutritional ideologies among part of the population. As a result, it is important to address several different aspects regarding the quality of meat and meat products, particularly those having to do with nutrition (as relates to health), safety and sustainability. Meat and meat products contain essential components of the diet, providing a large number of nutrients such as protein of high biological value, fat, mineral vitamins with high bioavailability, etc. However, its consumption has been linked to some negative health consequences due to some of its components such as lipids, salt, additives, and others. Consumers now view meat products as less healthy and less attractive, and this makes them more selective in the products they consume, as they are increasingly aware of improving their health through the foods they consume. Health is one of the main criteria for consumers when choosing the type of food they consume. In addition, this diet-health-disease relationship, together with a higher life expectancy, has sparked public administrations to promote the development of healthier foods with the aim of improving quality of life.

At present, there is growing market demand for healthier and more functional foods, and this is one of the main drivers of new trends and innovations in the development of new healthy products, including those of meat origin. Strategies for the development of healthy meat products are based mainly on the reduction or elimination of unhealthy components, which may or may not be replaced by a healthy alternative. This is done while maintaining the same quality criteria demanded of traditional products in terms of their sensory, technological, nutritional, functional, safety and other characteristics.

Strategies for the development of healthier meat products can be applied at different stages in the farm to fork food chain in order to obtain healthy and high-quality products. The first stage in the development of healthy meat products occurs at the farm during breeding and animal production through both genetic and nutritional strategies. Through these strategies, the composition of meat and its derivatives are modified and optimized with a healthier profile. However, it must be considered that any manipulation at this level, for example of animal diet, has effects on production and the final quality of the meat.

Another stage in the farm to fork food chain in which meat products with a more beneficial composition can be developed is the reformulation and elaboration of meat products. This is one of the most common technological strategies used in the design and development of new healthier meat products as it has an immediate impact on food composition as opposed to genetic modification or changes in animal feed. This strategy can be quickly applied by the meat industry. Regarding the reformulation of meat and meat products, different strategies have been explored to optimize the composition of these products, make them healthier and bring them in line with health recommendations and nutritional guidelines promoted by public bodies in response to new consumer demands [1]. There are basically two meat product reformulation strategies: reduce or eliminate certain unhealthy ingredients (salt, nitrites, saturated fatty acids, etc.) and add healthy ingredients (natural antioxidants, polyunsaturated fatty acids, probiotics, prebiotics, vitamins, etc.), or combinations of the two, the latter strategy being the most common. The beneficial compounds added to meat products can be of different origins (plant, marine, etc.). A current trend that is attracting considerable attention is the use of plant-based ingredients (extracts, by products, waste material, etc.), as multifunctional compounds in meat products. This is due to the added nutritional value and bioactive compounds (minerals, vitamins, polyunsaturated acids, antioxidants, etc.) and also to technological interest (water and fat binding capacity, emulsifying and gelling properties, etc.). Among the different reformulation strategies, considerable attention has been paid to those that optimize the lipid content and profile of meat products to meet nutritional needs and adhere to health recommendations. The excessive consumption of fatty foods is associated with health problems such as obesity, diabetes and cardiovascular disease that are all very common in today's society. The new lipid optimization strategies for the production of meat products seek to find or prepare ingredients similar to fat, such as analogues, imitators, mimetics, etc. One of today's most popular trends is lipid structuring (organogels, bulking agents, structured emulsions, etc.) [1]. Strategies aimed at reducing ingredients such as salt or additives such as nitrates and nitrites have also been thoroughly studied.

Meat product processing, conservation and consumption conditions are also considered critical stages in the process of gaining or losing healthy properties since, to a greater or lesser extent, these factors will alter products' components either by modifying or forming new compounds that may have positive or negative effects. This is the case, for example, of domestic practices once the product has been purchased. It is important to realize that once a healthier meat product has been developed, these beneficial properties must be maintained (farm to fork) so that it reaches the final consumer in optimal conditions and can have the beneficial effects intended.

The title of this Special Issue (SI) "Novel Strategies for the Development of Healthier Meat and Meat Products and Determination of Their Quality Characteristics" is based on all these considerations, our aim being to collect all new relevant information, which is of interest to all stakeholders (public institutions, livestock farmers, meat industry, technologists, researchers, teachers, students, etc.), and also to encourage researchers to search for new strategies to obtain healthier meat and meat products.

This special issue brings together 11 scientific contributions (2 reviews and 9 original research articles). It provides an interesting overview of the different strategies that can be applied at all stages from farm to fork; strategies that entail modifying meat and meat products to make them healthier and more sustainable. The SI also includes different meat and meat product quality control methods that have been developed.

Regarding strategies applied at the farm stage, a contribution on dietary manipulation is included, which studies the effects of algae meal supplements in feedlot lambs with competent reticular groove reflex (RGR) on growth performance, carcass traits and meat characteristics [2]. One of the main objectives of this strategy is to improve the quality of lamb meat by increasing omega-3 polyunsaturated fatty acid (PFA) levels. This is very interesting for the production of different meat products and for direct consumption as both are sources of omega-3 PFAs. In addition to this nutritional enhancement, the results of this study show that the addition of 2.5% seaweed to the diet of feedlot lambs with competent RGR has no detrimental effect on animal performance, carcass traits or quality characteristics of the meat (tenderness, colour, etc.).

Feng et al. [3] studied the quality characteristics (myofibril fragmentation index, total protein solubility, sarcoplasmic protein solubility, myofibrillar protein solubility, microstructure etc.) of different beef muscles (*semitendinosus, longissimus horacis, rhomboideus, gastrocnemius, infraspinatus, psoas major* and *biceps femoris*) during post-mortem ageing periods. The aim was to contribute new information to help improve and optimize the processing and storage of different cuts of beef in order to improve meat quality, mainly in terms of tenderness and acceptability. Both of these characteristics are critical indicators impacting people's willingness to purchase meat, and are also an issue in the elaboration of meat products containing these muscles.

Several of the articles in this SI focus on reformulation strategies. For example, Mahachi et al. [4] studied the use of different types of fat (fat-tailed sheep tail and pork back fat) in the development of novel warthog cabanossi. The study mainly focused on the physicochemical, fatty acid and sensory attributes of warthog cabanossi. This study concluded that the combination of fat-tailed sheep 'tail together with pork back fat could be an alternative to using only pork back fat to produce another variety of cabanossi of acceptable quality and the characteristic aroma and flavour but with some beneficial nutritional properties in terms of the n-3: n-6 ratio. This new product also gives consumers more options.

On the topic of reformulation, other authors have presented the use of mushrooms in the preparation of frankfurter [5] and nuggets [6] to reduce salt and fat in processed products and improve their nutritional profile. Both works focus on the use of plant by-products as functional food ingredients in meat products as these are a rich source of dietary fibre and various other bioactive compounds such as vitamins, minerals and polyphenols with strong antioxidant potential. Ceron-Guevara et al. [5] evaluated the effect that partially replacing fat and salt with edible mushroom flour (*Agaricus bisporus* and *Pleurotus ostreatus*) in frankfurter-type sausages had on physicochemical, microbiological and sensory parameters during cold storage. This reformulation strategy using edible mushroom flour is presented as an interesting substitute for fat and salt and even for meat. It improves the nutritional profile of the sausage, and the physicochemical and sensory properties were affected by the fungus, which gave it a strong umami flavour. Banerjee et al., (2020) [6] evaluated the impact of different amounts (2%, 4% and 6%) of enoki (*Flammulina velutipes*) mushroom stem waste powder on colour, texture, oxidative stability, sensory attributes and the shelf-life of goat meat nuggets. Based on the results of this study, Enoki stem waste extract can be used as a functional ingredient to improve the nutritional profile, physicochemical properties and shelf-life of meat products.

Following this same approach, Madane et al. [7] studied the efficacy of Moringa flower plant extract (MF) to develop a functional chicken product. The incorporation of this extract into chicken meat nuggets improved cooking performance and dietary fibre content without affecting the acceptability of the meat product and increased the lipid stability, odour score and shelf-life of chicken nuggets during refrigerated storage. The authors thus conclude that the MF extract could be used as a safe and natural antioxidant for the meat industry, which also offers functional health-promoting benefits.

Graso et al. [8] tested different concentrations (2 and 4%) of a potential sunflower seed by-product (obtained from sunflower oil extraction) as a substitute for animal fat to develop healthier sausages. The incorporation of this by-product led to a significant increase in protein, minerals (magnesium, potassium, copper, and manganese) and phenolic compounds, and lowered fat content (~37% less than the control made with animal fat). These results show that the use of this ingredient is a viable strategy to enhance the utility of sunflower oil by-products and obtain healthier sausages.

Franco et al. [9] also studied the substitution of pork fat in frankfurters, but in this case used linseed oleogel as the reformulation strategy. This oleogel substantially improved the fatty acid profile, SFA content and the n-6/n-3 ratio, although sensory and lipid oxidation parameters were significantly modified pointing to the need for further study.

Pintado et al. [10] presented a dual study. They first reduced fat by reformulating fresh sausages (longanizas) with chia (*Salvia hispanica* L.) or oat (*Avena sativa* L.) emulsion gel as animal fat replacers. They then studied the effect that cooking procedures (grilling) had on the composition and technological properties of longanizas to determine how nutritional and health claims were affected by this cooking process. After cooking, longanizas prepared with emulsion gels maintained their composition, allowing them to make nutritional and health claims in accordance with European legislation.

Lastly, this SI features two reviews. One [11] presents an interesting bibliographic search on fruit extracts and agro-industrial waste [Jabuticaba (*Myrciaria cauliflora*), Grape (*Vitis* sp.) and Nopal (*Opuntia ficus-indica*)] as sources of compounds potentially acting as antimicrobial agents. These extracts have an important application in meat products as they improve stability and are an interesting alternative to synthetic preservatives. The reduction or replacement of the more traditional preservatives that can be achieved with these extracts also provides products with cleaner labels, a growing trend in the reformulation of healthier compounds where sensory and technological properties remain intact without compromising safety. Use of these by-products also reduces environmental impact. Hence, the search for plant extracts with antimicrobial activity is an area of study on the rise for use in all types of food. The other review by Ruiz-Capillas and Herrero [1] stresses the importance of developing healthier meat products at all stages of the food chain (products, industry, consumers, administration, health organizations, etc.). It focuses mainly on reformulation in relation to healthier lipid content and lipid profile as this is one of the food components that has received the most attention. It also studies the different strategies used to that end such as the novel strategy of forming lipid materials based on structured lipids such emulsion gels (EGs) or oil-bulking agents (OBAs) that offer attractive applications in the reformulation of health-enhanced meat products. The review also conducts a critical analysis of the use of vibrational spectroscopy as a tool to further these developments. This interesting tool helps us gain a better understanding of the structural characteristics of lipid materials (EGs or OBAs) and of the corresponding reformulated health-enhanced meat products into which these fat replacers have been incorporated. These insights help in selecting the best lipid material to achieve specific technological properties in healthier meat products with improved lipid content and acceptable characteristics.

Meat and meat products are extensively consumed throughout the world and make an important nutritional contribution to our diet. However, their consumption has also been associated with some negative health consequences due to some of their components. This has driven significant growth in the field of design and development of healthy meat products focusing mainly on improving their composition, which in turn improves the diet-health relationship of these products. Different strategies have been used at different stages of the farm to fork journey to produce healthier products based on various compounds offering health benefits and featuring appropriate and safe technological and organoleptic properties, which are as similar to those of the original meat product as possible. It should also be noted that any strategy used to produce new healthy meat products will have an effect on the final product, which must be determined and controlled in different ways.

Today there is a great need to swiftly develop heathier meat and meat products to satisfy the demands of consumers who are increasingly aware of the link between diet and health and to comply with health recommendations made by different agencies. Science also needs to be a part of industrial processing so that scientific research can impact consumers in the form of optimal quality products.

The different strategies included in this SI are of interest to all those involved in meat and meat product industry from farm to fork. Lastly, many of the studies featured in this SI leave the door open to future research in this area of growing interest, not only for the meat and meat product industry, but for other foods as well.

Author Contributions: C.R.-C. and A.M.H.: writing—original draft preparation; C.R.-C. and A.M.H.: writing—review and editing; C.R.-C. and A.M.H.: funding acquisition. All authors have read and agreed to the published version of the manuscript.

Funding: This research was funded by the Spanish Ministry of Science and Innovation (PID2019-107542RB-C21), by the CSIC Intramural projects (grant numbers 201470E073 and 202070E242), CYTED (Reference 119RT0568; Healthy Meat network).

Acknowledgments: We thank to all the authors who have collaborated and have made this special issue possible.

Conflicts of Interest: The authors declare no conflict of interest.

References

1. Ruiz-Capillas, C.; Herrero, A.M. Development of Meat Products with Healthier Lipid Content: Vibrational Spectroscopy. *Foods* **2021**, *10*, 341. [CrossRef]
2. Núñez-Sánchez, N.; Ramírez, C.A.; Blanco, F.P.; Gómez-Cortés, P.; De La Fuente, M.Á.; Amor, M.V.; Ibáñez, A.H.; Marín, A.L.M. Effects of Algae Meal Supplementation in Feedlot Lambs with Competent Reticular Groove Reflex on Growth Performance, Carcass Traits and Meat Characteristics. *Foods* **2021**, *10*, 857. [CrossRef] [PubMed]
3. Feng, Y.H.; Zhang, S.S.; Sun, B.Z.; Xie, P.; Wen, K.X.; Xu, C.C. Changes in Physical Meat Traits, Protein Solubility, and the Microstructure of Different Beef Muscles during Post-Mortem Aging. *Foods* **2020**, *9*, 806. [CrossRef] [PubMed]
4. Mahachi, L.N.; Rudman, M.; Arnaud, E.; Muchenje, V.; Hoffman, L.C. Application of Fat-Tailed Sheep Tail and Backfat to Develop Novel Warthog Cabanossi with Distinct Sensory Attributes. *Foods* **2020**, *9*, 1822. [CrossRef] [PubMed]
5. Cerón-Guevara, M.I.; Rangel-Vargas, E.; Lorenzo, J.M.; Bermúdez, R.; Pateiro, M.; Rodríguez, J.A.; Sánchez-Ortega, I.; Santos, E.M. Reduction of Salt and Fat in Frankfurter Sausages by Addition of *Agaricus bisporus* and *Pleurotus ostreatus* Flour. *Foods* **2020**, *9*, 760. [CrossRef]
6. Banerjee, D.K.; Das, A.K.; Banerjee, R.; Pateiro, M.; Nanda, P.K.; Gadekar, Y.P.; Biswas, S.; McClements, D.J.; Lorenzo, J.M. Application of Enoki Mushroom (*Flammulina velutipes*) Stem Wastes as Functional Ingredients in Goat Meat Nuggets. *Foods* **2020**, *9*, 432. [CrossRef]
7. Madane, P.; Das, A.K.; Pateiro, M.; Nanda, P.K.; Bandyopadhyay, S.; Jagtap, P.; Barba, F.J.; Shewalkar, A.; Maity, B.; Jose, M.; et al. Drumstick (*Moringa oleifera*) Flower as an Antioxidant Dietary Fibre in Chicken Meat Nuggets. *Foods* **2019**, *8*, 307. [CrossRef] [PubMed]
8. Grasso, S.; Pintado, T.; Pérez-Jiménez, J.; Ruiz-Capillas, C.; Herrero, A.H. Potential of a Sunflower Seed By-Product as Animal Fat Replacer in Healthier Frankfurters. *Foods* **2020**, *9*, 445. [CrossRef] [PubMed]
9. Franco, D.; Martins, A.J.; López-Pedrouso, M.; Purriños, L.; Cerqueira, M.A.; Vicente, A.A.; Pastrana, L.M.; Zapata, C.; Lorenzo, J.M. Strategy towards Replacing Pork Backfat with a Linseed Oleogel in Frankfurter Sausages and Its Evaluation on Physicochemical, Nutritional, and Sensory Characteristics. *Foods* **2019**, *8*, 366. [CrossRef] [PubMed]
10. Pintado, T.; Ruiz-Capillas, C.; Jiménez-Colmenero, F.; Herrero, A.M. Impact of Culinary Procedures on Nutritional and Technological Properties of Reduced-Fat Longanizas Formulated with Chia (*Salvia hispanica* L.) or Oat (*Avena sativa* L.) Emulsion Gel. *Foods* **2020**, *9*, 1847. [CrossRef] [PubMed]
11. Gonçalves, L.A.; Lorenzo, J.M.; Trindade, M.A. Fruit and Agro-Industrial Waste Extracts as Potential Antimicrobials in Meat Products: A Brief Review. *Foods* **2021**, *10*, 1469. [CrossRef] [PubMed]

Article

Effects of Algae Meal Supplementation in Feedlot Lambs with Competent Reticular Groove Reflex on Growth Performance, Carcass Traits and Meat Characteristics

Nieves Núñez-Sánchez [1], Carmen Avilés Ramírez [2], Francisco Peña Blanco [1], Pilar Gómez-Cortés [3], Miguel Ángel de la Fuente [3], Montserrat Vioque Amor [2], Alberto Horcada Ibáñez [4] and Andrés Luis Martínez Marín [1],*

1   Departamento de Producción Animal, Universidad de Córdoba, Ctra. Madrid-Cádiz km 396, 14071 Córdoba, Spain; pa2nusan@uco.es (N.N.-S.); pa1peblf@uco.es (F.P.B.)
2   Departamento de Bromatología y Tecnología de los Alimentos, Universidad de Córdoba, Ctra. Madrid-Cádiz km 396, 14071 Córdoba, Spain; v92avrac@uco.es (C.A.R.); bt1viamm@uco.es (M.V.A.)
3   Instituto de Investigación en Ciencias de la Alimentación (CIAL, CSIC-UAM), Universidad Autónoma de Madrid, Nicolás Cabrera, 9, 28049 Madrid, Spain; p.g.cortes@csic.es (P.G.-C.); mafl@if.csic.es (M.Á.d.l.F.)
4   Departamento de Agronomía, Escuela Técnica Superior de Ingeniería Agronómica, Universidad de Sevilla, 41013 Sevilla, Spain; albertohi@us.es
*   Correspondence: pa1martm@uco.es; Tel.: +34-957-218-746

Citation: Núñez-Sánchez, N.; Avilés Ramírez, C.; Peña Blanco, F.; Gómez-Cortés, P.; de la Fuente, M.Á.; Vioque Amor, M.; Horcada Ibáñez, A.; Martínez Marín, A.L. Effects of Algae Meal Supplementation in Feedlot Lambs with Competent Reticular Groove Reflex on Growth Performance, Carcass Traits and Meat Characteristics. *Foods* **2021**, *10*, 857. https://doi.org/10.3390/foods10040857

Academic Editors: Claudia Ruiz-Capillas, Ana Herrero Herranz and Mohammed Gagaoua

Received: 18 March 2021
Accepted: 13 April 2021
Published: 15 April 2021

Publisher's Note: MDPI stays neutral with regard to jurisdictional claims in published maps and institutional affiliations.

Abstract: There is growing interest in increasing omega-3 fatty acid (FA) contents in ruminant meat by means of dietary manipulation, but the effects of such manipulation on productive results and meat quality need to be ascertained. The aim of the present study was to assess the effects of supplementing lambs with competent reticular groove reflex (RGR) with marine algae as a source of omega-3 fatty acids on growth performance, carcass traits, and meat quality characteristics. Forty-eight feedlot lambs were distributed into three equal groups: the control group neither consumed marine algae nor had competent RGR, the second group received daily 2.5% of algae meal mixed in the concentrate, and the last group consumed the same amount of algae meal, but emulsified in a milk replacer and bottle-fed. Lambs in the second and third groups had competent RGR. There were not any negative effects on performance, carcass or meat quality parameters with algae supplementation. However, the results of the oxidative stability parameters were not conclusive. Ageing for 6 days improved meat tenderness and color, and increased lipid oxidation. In conclusion, algae meal inclusion in the diet of fattening lambs with competent RGR has no detrimental effects on animal performance, carcass traits or meat quality characteristics.

Keywords: marine algae; lambs; performance; meat quality

1. Introduction

Lamb meat is widely consumed in some geographical areas, for instance in Mediterranean countries, but has a non-healthy nutritional image, mostly due to the general idea of having high levels of saturated fatty acids (FA), variable contents of trans-fat, and low levels of omega-3 polyunsaturated FA [1]. However, the quality of lamb meat and its FA profile is closely related to the feeding conditions of the animals. The intramuscular fat (IMF) of lambs reared under intensive feeding conditions is characterized by high levels of saturated and omega-6 FA, and a low amount of omega-3 FA [2–5]. In contrast, meat from grass-fed lambs has shown a more desirable FA composition, with lower contents of saturated FA and higher levels of omega-3 FA [6].

In the last decade, there has been growing interest in finding appropriate and natural ways to manipulate IMF composition, with grazing being one of the best alternatives. However, in regions where fattening on pasture is not feasible for climatic reasons, other approaches to increase omega-3 FA have been evaluated, such as adding plant-derived oils rich in omega-3 FA [7–9]. Dietary marine algae have been shown to improve meat nutritional value in lambs [8,10–13] due to their high eicosapentaenoic (EPA), docosapentaenoic (DPA) and docosahexaenoic (DHA) FA contents. Very recently, our group has reported that the dietary inclusion of marine algae at the level of 2.5% enhanced omega-3 FA contents in lamb meat with competent reticular groove reflex (RGR), with the increase being significantly higher when the marine algae was bottle-fed in comparison to the same amount of algae meal mixed in the concentrate [14]. Fostering the RGR of the newborn animal into adulthood and using it to include emulsified lipid sources into the abomasum, bypassing the rumen, has been demonstrated as a fruitful strategy to enhance the healthy unsaturated FA presence in ruminant milk and meat [14,15].

Marine algae supplementation may have undesirable effects on growth performance, carcass traits and meat quality characteristics [16], such as a decrease in average daily gain and feed intake, and thus an extension of the fattening period [10,17]. Moreover, the high content of polyunsaturated FA in the IMF from lambs fed marine algae increases its peroxidability index [14], which could lead to meat quality degradation during storage [18]. Changes occurring in the meat micro- and ultra-structure during the aging process are associated with favorable modifications in its meat tenderization and water-holding capacity. However, as stated previously, the oxidation of lipid components, as well as the destabilization of meat color, can negatively affect the meat's final value [19]. Again, marine algae are a good source of antioxidants. Most researchers consider polyphenolic compounds (i.e., phenolic and cinnamic acids, phlorotannins, and bromophenols) among the main factors responsible for such antioxidant properties [20].

The farm to fork strategy includes hygienic, compositional, nutritional, sensory, and technological quality characteristics in the general concept of meat quality throughout the food chain, in order to obtain high-quality products [21].

The aim of this study was to assess the effects of supplementing lambs with competent RGR with marine algae as a source of omega-3 FA on growth performance, carcass traits and meat quality characteristics.

2. Materials and Methods

2.1. Experimental Design and Diets

This experiment was carried out on the premises of the Animal Production building of the University of Córdoba (Spain). Details about the experimental design and diets have recently been published [14]. Briefly, a total of 48 male Manchega-breed lambs at 42 days of age were weighed (11.6 ± 1.67 kg) and assigned in pairs of similar body weights to 1 of 24 adjacent pens (1.40 m² raised slatted floor cages with individual troughs for feed and water, within an environmental controlled room). Pens were distributed in 8 blocks according to average body weight and allocated randomly to one of the three treatments (16 animals per treatment): (i) the control group, which consisted of a typical pelleted concentrate without algae meal supplementation (NOALG), (ii) the algae meal in concentrate group, which received the same concentrate as NOALG but mixed with 2.5 % algae meal (*Aurantiochytrium limacinum*; Forplus, Alltech Spain, Guadalajara, Spain) plus 250 mL daily of milk replacer in a single feeding (ALGCON), and (iii) the algae meal in milk replacer group, which received the same concentrate as NOALG plus 250 mL daily of milk replacer supplemented with algae meal in a single feeding (ALGMILK). The concentrate was composed by 70% cereals (barley, maize and wheat, 40:20:10), 20% soybean meal, 6% wheat bran and 4% minerals and vitamins, so its calculated composition was 16% crude protein and 11.0 MJ/kg metabolizable energy, as fed. In the ALGCON and ALGMILK diets, milk replacer provided an extra 0.89 MJ of metabolizable energy per day, and algae meal a total of 0.75 MJ/kg of concentrate. Besides this, these diets caused an increase in the crude protein intake of 10 g/d with the milk replacer and 3 g/kg of concentrate with the algae meal. The ALGCON and ALGMILK treatments provided ~584 mg of omega-3 FA per 100 g of concentrate consumed. Lambs in the ALGCON and ALGMILK groups had competent RGR. All animals had free access to wheat straw and fresh water throughout the whole experimental period (49 days on fattening).

2.2. Sampling and Analysis

The average body weight per pen was recorded weekly during the experiment, in the morning before feeding. The average daily feed intake (ADFI, kg/day) per pen was calculated daily by weighing each day the amount of concentrate offered and subtracting the amount of feed refusals found in the feeders the following morning. Besides this, average daily gain (ADG, kg/day) and feed conversion ratio (FCR, feed to gain kg/kg) were calculated for each pen.

When the average body weight of the set of lambs reached ~25 kg (in the range of the commercial slaughter weight for this type of lambs), the animals were tagged to track their carcasses, and transported from the Animal Production facilities at the University of Cordoba to a commercial slaughterhouse (COVAP, Pozoblanco, Córdoba, Spain), located ~90 km (~1.0 h) away, in a vehicle adequately conditioned. Then, the animals were placed in pens (8 lambs per pen) and remained there for approximately 14 h, with free access to water but not feed. After lairage, lambs were stunned, slaughtered, and dressed.

The carcasses of the animals were weighed in the first 45 min after slaughter (hot carcass weight, HCW). The dressing was calculated as the ratio of HCW to final body weight and expressed as a percentage. Lamb carcasses were classified according to their weight as class A (<7 kg), class B (7.1 to 10 kg), and class C (10.1 to 13 kg). A trained and experienced technician from the slaughterhouse visually graded carcass fatness considering the size of the kidney and pelvic fat deposits (types 1, low fat; 2, medium fat; 3, high fat) and muscle color (pale pink, pink, red) [22].

After 2 h at room temperature, all carcasses were chilled at 4.0 °C for 24 h in a commercial chiller and transferred to the Animal Production laboratory without disrupting the cold chain. The left shoulders were weighed and dissected into muscle, bone and fat (subcutaneous, pre-scapular and intermuscular) and the remaining tissues (major blood vessels, ligaments, tendons and fascias). Each fraction was weighed, and the results were expressed as the percentage of total shoulder weight to provide an estimate of the carcass composition. At 24 h post-mortem, the *Longissimus thoracis* muscle (T6 to T13 vertebrae) from each left carcasses was removed and divided into three pieces (T6, T7 to T12, and T13 vertebrae), which were further used for the assessment of meat quality (pH, color, drip and cooking losses, and Warner–Bratzler shear force) and parameters related to meat oxidative stability. The *Longissimus thoracis* muscle from the right carcass was vacuum-packed, aged in a refrigerated chamber at 2–4 °C in the dark for 6 days (7 d postmortem), and further used for the same determinations.

The methodology of the measurements is described in detail in Avilés et al. [2]. Briefly, the drip loss (DL) of each sample was expressed as the percentage of weight loss of a sample hanging at 4° for 24 h, related to the initial weight. The pH was measured by inserting the glass electrode of a portable pH-meter (Crison® PH25, Hach Lange, Barcelona, Spain) approximately 1 cm into the *Longissimus thoracis* muscle between the T11 and T12 junction sites. Meat color was measured on the freshly cut surface of the middle section of the T12 vertebra, after 30 min of blooming at room temperature in the dark. A CM-2600d hand-held spectrophotometer (D65 illuminant, 8 mm diameter aperture, 10° standard observer, 8° viewing angle; Minolta Inc., Osaka, Japan) was used for color determination, according to the CIE system (CIE, 1986). Three measurements of color coordinates, expressed as L* (lightness), a* (redness) and b* (yellowness), were performed, and the average was used for the calculation of color saturation (C*) and hue (h°). For the determination of cooking loss (CL) and Warner–Braztler shear force (WBSF), the samples were weighed, cooked in a plastic bag in a water bath at 75 °C until the temperature in the center of the sample reached 70 °C (monitored by a HI 98509 Checktemp® Pocket Thermometer, Hanna Instruments, Guipuzcoa, Spain), and then cooled at room temperature for 30 min, blotted dry, and weighed again. CL was expressed as the percentage of weight loss related to the initial weight. WBSF values, expressed in kg/cm^2, were taken for muscle cores of 1 cm^2 cross-sections using a texture analyzer (TA.TXT-2, Stable Micro Systems, Surrey, UK) equipped with a Warner–Bratzler shearing device (Mitutoyo series 500, Mitutoyo Corporation, Aurora, IL, USA). A total of 10 measurements was performed in each meat sample, and the average was provided as the WBSF value of the sample. The extent of lipid oxidation was assessed by measuring the thiobarbituric acid reacting substances (TBARS) using the method of Tarladgis et al. [23], as described in Avilés et al. [2]. The 1,1,3,3-tetra-ethoxypropane (TEP) standard curve was used for calculating the TBARS concentration and the results were expressed as mg of MDA kg^{-1} of meat. The antioxidant capacity of the meat was determined through the 2,2-diphenylpicrylhydrazyl (DPPH) radical scavenging capacity assay. The meat hydrophilic and lipophilic extracts were obtained following the method described by Folch et al. [24]. The determination of the free radical scavenging activity of the hydrophilic and lipophilic extracts of the meat samples (DPPH water and DPPH fat, respectively) was carried out following the procedure described by Brand-Williams et al. [25]. The absorbance values in both hydrophilic and lipophilic extracts were expressed as % radical scavenging activity (RSA). Total polyphenols content was determined with the Folin–Ciocalteu reagent according to the procedure described by Singleton [26]. Total phenolic compounds were quantified using a gallic acid reference standard, and the results were expressed as mg of gallic acid equivalents (GAE)/100 g of meat.

2.3. Statistical Analysis

SAS UE 3.8 software (SAS Institute Inc., Cary, NC, USA) was used to perform the statistical analyses. Statistical significance was declared at $p < 0.05$. Productive and carcass traits, except for grading, were analyzed with the MIXED procedure. The statistical model included the treatment as fixed effect and the pen nested within the treatment as the random effect. When the fixed effect was significant, differences between least squares means were assessed by Tukeys's test. A repeated measurements analysis of meat data was carried out with the MIXED procedure. The statistical model included the fixed effects of treatment, ageing time, and their interaction; the repeated effect was ageing time; the subject of the repeated measurements was the animal nested within the treatment and the pen. When the fixed effects of the repeated measurement model were significant, differences between least squares means were assessed by paired *t*-test.

3. Results

3.1. Growth Performance and Carcass Traits

Growth performance was not affected ($p > 0.05$) by the addition of algae meal to the experimental rations (Table 1). According to HCW, 22% of the carcasses were classified as class A and 78% as class B, whereas according to fatness most carcasses were within type 2 (60% in comparison to 34% and 6% in types 3 and 1, respectively). The subjective carcass color was pale pink and pink in 65 and 35% of the carcasses, respectively. The lowest HCW and dressing percentage were observed ($p < 0.05$) in ALGMILK, while ALGCON treatment showed the highest values (Table 1). ALGCON treatment had lower and higher percentages of carcass muscle and fat, respectively, than NOALG ($p < 0.05$), while ALGMILK showed intermediate values.

Table 1. Growth performance and carcass characteristics of feedlot lambs fed a conventional diet alone (NOALG) or feedlot lambs with competent reticular groove reflex fed the same diet supplemented with 2.5% of algae meal, either mixed in the concentrate (ALGCON) or in the milk replacer (ALGMILK).

Parameters [1]	DIET			SEM	*p*
	NOALG	**ALGCON**	**ALGMILK**	**SEM**	*p*
Initial body weight (kg)	11.4	11.6	11.8	0.24	0.86
Final body weight (kg)	25.0	25.7	25.4	0.31	0.78
Average feed intake (g)	825	801	805	6.0	0.44
Average daily gain (g)	326	335	324	4.2	0.33
Feed conversion ratio (kg/kg)	2.54	2.43	2.49	0.030	0.13
Hot carcass weight (kg)	10.5 [ab]	11.4 [a]	10.4 [b]	0.16	<0.05
Dressing [2] (%)	42.2 [ab]	44.6 [a]	40.9 [b]	0.60	<0.05
Carcass composition [3]:					
Muscle (%)	60.8 [a]	58.0 [b]	59.8 [ab]	0.38	<0.05
Fat (%)	14.5 [b]	17.2 [a]	15.2 [ab]	0.39	<0.05
Bone (%)	24.8	24.8	24.9	0.19	0.95

[1] Within each treatment, 8 pens were used for determination of growth performance (2 animals per pen) and 16 animals were used for determination of carcass traits. [2] Calculated as hot carcass weight/final body weight. [3] Estimated by shoulder dissection. SEM: standard error of the mean. Means with different superscripts between treatments are significantly different ($p < 0.05$).

3.2. Meat Quality Characteristics and Oxidative Stability

The meat quality characteristics and oxidative stability results are shown in Table 2. Neither the treatment nor the ageing time affected meat pH or DL ($p > 0.05$). CL and WBSF parameters were not modified by the treatments ($p > 0.05$), but were reduced by ageing time ($p < 0.05$). The addition of algae meal to the diet did not affect meat color ($p > 0.05$), apart from the b* coordinate, which was lower ($p < 0.05$) in NOALG. On the contrary, ageing time influenced ($p < 0.05$) the color coordinates, causing a significant increase in L* and a*, as well as a decrease in the b*, C* and h° parameters in all experimental groups.

The presence of algae meal in the diet tended to increase TBARS ($p = 0.06$), while ageing increased ($p < 0.05$) TBARS. At day 1 of ageing, the DPPH fat was higher ($p < 0.05$) in the ALGCON than in the ALGMILK treatment, with an intermediate value for NOALG, but the differences disappeared ($p > 0.05$) at day 6 of ageing. DPPH water did not differ ($p > 0.05$) among treatments, but decreased ($p < 0.05$) with ageing time. As regards the polyphenol content in the diets, the algae meal supplementation supplied 11.8 mg GAE/100 g, while the amount of GAE/100 g in the NOALG diet was 10.3 mg. The total polyphenols contents in meat were not affected by treatments or ageing.

Regarding oxidative stability, it is worth mentioning that marine algae increased total SFA levels in the ALGCON and ALGMILK treatments (40.3 and 40.2%, respectively, vs. 36.9% of total FA in the NOALG treatment), did not affect total MUFA (38.15, 37.86, and 36.60% of total FA in the NOALG, ALGCON, and ALGMILK treatments, respectively), and raised total omega-3 FA (1.45, 3.94 and 4.96 % of total FA in the NOALG, ALGCON, and ALGMILK treatments, respectively). The increases in the levels of omega-3 FA, EPA, DPA, and, most significantly, DHA, were higher when the algae meal was bottle-fed via RGR, in comparison to algae meal mixed in the concentrate [14].

Table 2. Quality characteristics and oxidative stability of meat samples from feedlot lambs fed a conventional diet alone (NOALG) or from feedlot lambs with competent reticular groove reflex fed the same diet supplemented with 2.5% of algae meal, either mixed in the concentrate (ALGCON) or in the milk replacer (ALGMILK).

Parameters [1]	Ageing Time (A)	Diet (D)			SEM	p		
		NOALG	ALGCON	ALGMILK		D	A	D × A
Meat characteristics								
pH_{24}	1	5.77	5.70	5.73	0.01	0.21	0.19	0.87
	7	5.78	5.72	5.74				
Drip Loss (%)	1	1.93	1.74	1.63	0.06	0.46	0.24	0.06
	7	1.49	1.86	1.59				
Cooking Loss (%)	1	21.5	22.7	20.6	0.96	0.33	<0.01	0.19
	7	19.9	12.7	15.0				
WBSF (kg/cm^2)	1	7.85	7.28	7.28	0.21	0.97	<0.001	0.07
	7	4.85	5.18	5.38				
L*	1	37.4	38.7	38.8	0.34	0.30	<0.001	0.87
	7	41.6	42.7	42.6				
a*	1	6.27	5.67	6.06	0.20	0.97	<0.001	0.09
	7	7.87	8.55	8.29				
b*	1	16.4	16.7	16.7	0.32	<0.05	<0.001	0.18
	7	10.0	11.1	11.1				
C*	1	17.6	17.7	17.7	0.24	0.20	<0.001	0.13
	7	13.0	14.2	14.0				
h°	1	69.3	71.5	70.2	1.09	0.54	<0.001	0.44
	7	50.7	51.6	52.5				
Oxidative stability								
TBARS (mg MDA/kg)	1	0.23	0.67	0.62	0.05	0.06	<0.001	0.24
	7	0.97	1.06	1.14				
DPPH fat (%)	1	24.2 [AB,a]	29.0 [A,a]	17.5 [B]	1.03	<0.05	<0.001	<0.001
	7	14.6 [b]	15.0 [b]	14.4				
DPPH water (%)	1	6.28	6.76	6.63	0.31	0.78	<0.01	0.96
	7	4.20	4.58	4.81				
Polyphenols (mg GAE/100 g)	1	3.72	3.93	4.34	0.07	0.19	0.92	0.21
	7	4.10	3.91	4.03				

[1] In each time within each treatment, 16 samples were analyzed for each parameter. SEM: standard error of the mean. L*: lightness; a*: redness; b*: yellowness; C*: chroma; h°: hue; WBSF: Warner–Braztler shear force; TBARS: thiobarbituric acid reactive substances; DPPH fat: 2-2 diphenyl picryl hydrazyl in fat extract; DPPH water: 2-2 diphenyl picryl hydrazyl in aqueous extract. Means with different superscript letters between treatments (capital) or ageing times (lowercase) are significantly different.

4. Discussion

4.1. Growth Performance and Carcass Characteristics

The slaughter weights of lambs fed different experimental diets were within the range of 22–30 kg BW, as established in the European Regulation for "Cordero Manchego" protected geographical indication [27]. The average feed intake, average daily gain and feed conversion ratio values from the present study were more favorable than those obtained previously by Avilés et al. [2] using lambs from the same genetic background, with similar ages and slaughter weights, raised in on-farm conditions.

The overall performance of the lambs was not affected by the addition of algae meal to the diet in the present study, which is in agreement with previous research, wherein the diets of fattening lambs were supplemented with *Aurantiochytrium limacinum*, *Arthrospira platensis*, and *Isochiris* sp. [8,11,28,29]. However, the results in the literature regarding algae meal supplementation to the diet of lambs are somehow contradictory. Other researchers found that lambs fed diets containing *Aurantiochytrium limacinum* algae had slower growth rates at 1.7% of inclusion [7], or lower average feed intakes and daily gains at 2 to 6% of inclusion [10,13,17], in comparison with the control lambs. In contrast, increased average feed intake, daily gain and final body weight in lambs fed *Arthrospira platensis* algae meal at 1 g/10 kg BW/d were also reported [30]. These discrepancies could be mainly attributed to the type and amount of algae used, its palatability, the trial duration, and/or the weights and ages of the lambs following algae treatments.

As in the present study, previous research did not find differences in HCW and dressing percentage between control and algae-added diets [7,8,11,13,29]. Nonetheless, the means of administration of the algae meal showed a marked influence on HCW and dressing, being significantly higher when algae meal was fed in the concentrate (ALGCON) and reduced when the same amount of supplement was fed in the milk replacer (ALGMILK). Regarding the carcass composition, the present research is in agreement with previous results [2], and the reported values are within the normal ranges for this type of meat. On average, the algae meal treatments showed ~12% more carcass fat than NOALG, which might be related to a higher percentage of fat in the former's daily gain due to the extra energy consumed via milk replacer and algae meal [31].

4.2. Meat Quality Characteristics and Oxidative Stability

Ultimate pH (24 h after slaughter) and its fall rate are widely used to evaluate raw meat quality, due to its strong relationship with meat quality characteristics such as color, water-holding capacity and tenderness [32,33]. In the present study, neither treatments nor ageing time affected meat pH (Table 2). The ultimate pH ranged from 5.70 to 5.77 (Table 2), values typically observed in non-stressed sheep at the time of slaughter [34,35]. The observed pH values (Table 2) were within the expected range for this type of lamb meat, and were in agreement with those reported by other authors [2,8,17,29]. The lack of changes in meat pH due to ageing is in line with previous studies [19,36,37].

DL was not influenced by the addition of algae meal to the diet, and remained stable throughout ageing (Table 2). The DL values measured at day 1 were in agreement with those reported by other authors [8,13]. Regarding aged samples, the DL values were slightly higher than those shown by Avilés et al. [2] and Vergara et al. [35,37] in Manchega lambs.

Treatments had no effect on CL, or loss of water with cooking. The CL values (Table 2) differed from those obtained by other authors [2,38,39]. The slaughter weight, fatness, pH, cooking procedure and cooling time, among others, are factors to which these differences could be attributed [40]. The numerical trend of lower CL in day 7 samples matched with the observed trend of carcass fat percentage, suggesting that as fat increased, the aged samples were protected from losing water [41,42]. Conversely, Hopkins et al. [29] found that the CL values were significantly higher in algae-fed lambs, but those animals had no increased subcutaneous fat depth.

Meat tenderness, measured as WBSF, was not affected by the inclusion of algae meal in the diet, neither in fresh nor in aged meat samples, which is in agreement with Valença et al. [13] and Hopkins et al. [29]. The average WBSF value at day 1 was greater than the threshold (5 kg/cm^2) reported by Shorthose et al. [43] to classify lamb meat into tender or tough, while, after 6 days of ageing, it was similar to that value. The maximum shear force recorded in the present study was similar to that obtained by Avilés et al. [2], Blanco et al. [38] and Linares et al. [44]. As expected, ageing time caused a significant decrease in WBSF, as previously described [45].

The meat color coordinates L*, a*, C* and h° were not affected by the algae meal treatments (Table 2), in agreement with previous research [8,12,13,29]. Although b* was higher on average in the algae meal treatments, it did not affect h° and C* values, and thus did not alter the perception of color (from red to yellow) or its vividness, respectively [46]. Desirable changes in color indices were observed during ageing in the present study (i.e., higher L* and a* values and lower b* and h° values), as observed by other authors [39].

Algae meal in the diet tended to increase meat oxidation measured as TBARS in day 1 samples (Table 2), whereas ageing significantly raised TBARS regardless of the treatment. All TBARS values were lower than the acceptability limit of 2 mg MDA kg^{-1} muscle to detect rancidity or oxidized flavors in cooked lamb meat by consumers [18,47,48]. Previous research has shown that algae meal supplementation increases meat lipid oxidation levels [8,12,13,17], which can be attributed to the higher long chain omega-3 polyunsaturated FA content in the IMF of algae meal-supplemented lambs [14].

Meat antioxidant activity was evaluated through the DPPH radical scavenging capacity of both lipophilic (DPPH fat) and hydrophilic (DPPH water) extracts (Table 2). Day1 ALGMILK samples, which contained 5 g of omega-3 FA/100 g of total fat [14], showed less DPPH fat than those from ALGCON treatment, which contained 20% less omega-3 FA. This difference would be mainly related to the need to eliminate the free radicals generated by the extra omega-3 FA in ALGMILK samples [18]. Ageing time decreased both the DPPH fat and the DPPH water of NOALG and ALGCON samples, but a similar antioxidant capacity was observed in all aged meats.

The meat's total polyphenols contents were not affected by dietary treatments (Table 2), which is in agreement with Muiño et al. [49], who did not observe differences after the supplementation of sheep diets with red wine polyphenols (900 mg of red wine extract/kg of feed). On the contrary, Luciano et al. [50] reported higher concentrations of polyphenols in the meat of lambs fed a diet enriched with tannins for 60 days (8.96% quebracho supplement rich in proanthocyanidins), compared to control lambs. Although algae are recognized as a good source of polyphenols, it may have been the case that the amounts of marine algae included in the current experimental diets or the length of supplementation were not sufficient to trigger any significant response [51].

5. Conclusions

The results of this study show that the addition of 2.5% marine algae to the diet of fattening lambs with competent RGR, either mixed in the concentrate or bottle-fed, does not have any negative effect on growth performance, carcass characteristics, meat quality, or, for the most part, oxidative stability. This "from farm to fork" strategy, intended to increase the levels of omega-3 polyunsaturated FA in meat, would be of great interest to the improvement of lamb meat from a nutritional point of view, without affecting animal performance or other meat quality parameters.

Author Contributions: Conceptualization, F.P.B. and A.L.M.M.; methodology, F.P.B., M.V.A., C.A.R. and A.H.I.; formal analysis, A.L.M.M.; investigation, N.N.-S., F.P.B. and A.L.M.M.; resources, A.L.M.M., F.P.B., M.V.A. and A.H.I.; data curation, F.P.B. and C.A.R.; writing—original draft preparation, N.N.-S., F.P.B. and C.A.R.; writing—review and editing, A.L.M.M., P.G.-C. and N.N.-S.; visualization, P.G.-C.; supervision, A.L.M.M., P.G.-C. and M.Á.d.l.F.; project administration, M.Á.d.l.F.; funding acquisition, M.Á.d.l.F. All authors have read and agreed to the published version of the manuscript.

Funding: This research was funded by the Spanish Ministry of Science and Innovation, project number AGL2016-75159-C2-2-R and the Spanish National Research Council, grant number 20217AT002.

Institutional Review Board Statement: The study was approved by the Ethics Committee of Universidad de Córdoba and the competent authorities (protocol code 04/05/2018/074 and date 5 May 2018).

Informed Consent Statement: Not applicable.

Data Availability Statement: The data presented in this study are available in the article.

Acknowledgments: P.G.-C. gratefully acknowledges the Ramón y Cajal research contract (RYC2019-027933-I) funded by the Spanish Ministry of Science and Innovation.

Conflicts of Interest: The authors declare no conflict of interest. The funders had no role in the design of the study; in the collection, analyses, or interpretation of data; in the writing of the manuscript, or in the decision to publish the results.

References

1. McAfee, A.J.; McSorley, E.M.; Cuskelly, G.J.; Moss, B.W.; Wallace, J.M.W.; Bonham, M.P.; Fearon, A.M. Red meat consumption: An overview of the risks and benefits. *Meat Sci.* **2010**, *84*, 1–13. [CrossRef] [PubMed]
2. Avilés Ramírez, C.; Peña Blanco, F.; Horcada Ibáñez, A.; Núñez Sánchez, N.; Requena Domenech, F.; Guzmán Medina, P.; Martínez Marín, A.L. Effects of concentrates rich in by-products on growth performance, carcass characteristics and meat quality traits of light lambs. *Anim. Prod. Sci.* **2019**, *59*, 593–599. [CrossRef]
3. Cooper, S.L.; Sinclair, L.A.; Wilkinson, R.G.; Hallett, K.G.; Enser, M.; Wood, J.D. Manipulation of the n-3 polyunsaturated fatty acid content of muscle and adipose tissue in lambs. *J. Anim. Sci.* **2004**, *82*, 1461–1470. [CrossRef] [PubMed]
4. Díaz, M.T.; Álvarez, I.; De la Fuente, J.; Sañudo, C.; Campo, M.M.; Oliver, M.A.; Font i Furnols, M.; Montossi, F.; San Julián, R.; Nute, G.R.; et al. Fatty acid composition of meat from typical lamb production systems of Spain, United Kingdom, Germany and Uruguay. *Meat Sci.* **2005**, *71*, 256–263. [CrossRef]
5. Ponnampalam, E.N.; Butler, K.L.; Pearce, K.M.; Mortimer, S.I.; Pethick, D.W.; Ball, A.J.; Hopkins, D.L. Sources of variation of health claimable long chain omega-3 fatty Australian lamb slaughtered at similar weights. *Meat Sci.* **2014**, *96*, 1095–1103. [CrossRef]
6. Elizalde, F.; Hepp, C.; Reyes, C.; Tapia, M.; Lira, R.; Morales, R.; Sales, F.; Catrileo, A.; Silva, M. Growth, Carcass and Meat Characteristics of grass-fed lambs weaned from extensive rangeland and grazed on permanent pastures or alfalfa. *Animals* **2021**, *11*, 52. [CrossRef]
7. Burnett, V.F.; Jacobs, J.L.; Norng, S.; Ponnampalam, E.N. Feed intake, liveweight gain and carcass traits of lambs offered pelleted annual pasture hay supplemented with flaxseed (*Linum usitatissimum*) flakes or algae (*Schizochytrium* sp.). *Anim. Prod. Sci.* **2017**, *57*, 877–883. [CrossRef]
8. De la Fuente-Vázquez, J.; Díaz-Chirón, M.T.; Pérez-Marcos, C.; Cañeque-Martínez, V.; Sánchez-González, C.I.; Álvarez-Acero, I.; Fernández-Bermejo, C.; Rivas-Cañedo, A.; Lauzurica-Gómez, S. Linseed, microalgae or fish oil dietary supplementation affects performance and quality characteristics of light lambs. *Span. J. Agric. Res.* **2014**, *12*, 436–447. [CrossRef]
9. Nguyen, D.V.; Malau-Aduli, B.S.; Cavalieri, J.; Nichols, P.D.; Malau-Aduli, A.E.O. Supplementation with plant-derived oils rich in omega-3 polyunsaturated fatty acids for lamb production. *Vet. Anim. Sci.* **2018**, *6*, 29–40. [CrossRef]
10. Díaz, M.T.; Pérez, C.; Sánchez, C.I.; Lauzurica, S.; Cañeque, V.; González, C.; De La Fuente, J. Feeding microalgae increases omega 3 fatty acids of fat deposits and muscles in light lambs. *J. Food Compos. Anal.* **2017**, *56*, 115–123. [CrossRef]
11. Meale, S.J.; Chaves, A.V.; He, M.L.; McAllister, T.A. Dose-response of supplementing marine algae (*Schizochytrium* sp.) on production performance, fatty acid profiles, and wool parameters of growing lambs. *J. Anim. Sci.* **2014**, *92*, 2202–2213. [CrossRef]
12. Ponnampalam, E.N.; Burnett, V.F.; Norng, S.; Hopkins, D.L.; Plozza, T.; Jacobs, J.L. Muscle antioxidant (vitamin E) and major fatty acid groups, lipid oxidation and retail colour of meat from lambs fed a roughage based diet with flaxseed or algae. *Meat Sci.* **2016**, *111*, 154–160. [CrossRef]
13. De Lima Valença, R.; da Silva Sobrinho, A.G.; Borghi, T.H.; Meza, D.A.R.; de Andrade, N.; Silva, L.G.; Bezerra, L.R. Performance, carcass traits, physicochemical properties and fatty acids composition of lamb's meat fed diets with marine microalgae meal (*Schizochytrium* sp.). *Livest. Sci.* **2021**, *243*, 104387. [CrossRef]
14. Gómez-Cortés, P.; de la Fuente, M.A.; Peña Blanco, F.; Nuñez-Sánchez, N.; Requena Domenech, F.; Martínez Marín, A.L. Feeding algae meal to feedlot lambs with competent reticular groove reflex increases omega-3 fatty acids in meat. *Foods* **2021**, *10*, 366. [CrossRef]
15. Dobarganes García, C.; Pérez Hernández, M.; Cantalapiedra, G.; Salas, J.M.; Merino, J.A. Bypassing the rumen in dairy ewes: The reticular groove reflex vs. calcium soap of olive fatty acids. *J. Dairy Sci.* **2005**, *88*, 741–747. [CrossRef]
16. Madeira, M.S.; Cardoso, C.; Lopes, P.A.; Coelho, D.; Afonso, C.; Bandarra, N.M.; Prates, J.A.M. Microalgae as feed ingredients for livestock production and meat quality: A review. *Livest. Sci.* **2017**, *205*, 111–121. [CrossRef]
17. Urrutia, O.; Mendizabal, J.A.; Insausti, K.; Soret, B.; Purroy, A.; Arana, A. Effects of addition of linseed and marine algae to the diet on adipose tissue development, fatty acid profile, lipogenic gene expression, and meat quality in lambs. *PLoS ONE* **2016**, *11*. [CrossRef] [PubMed]
18. Ponnampalam, E.N.; Butler, K.L.; Muir, S.K.; Plozza, T.E.; Kerr, M.G.; Brown, W.G.; Jacobs, J.L.; Knight, M.I. Lipid oxidation and colour stability of lamb and yearling meat (muscle *Longissimus lumborum*) from sheep supplemented with camelina-based diets after short-, medium-, and long-term storage. *Antioxidants* **2021**, *10*, 166. [CrossRef]
19. Rant, W.; Radzik-Rant, A.; Świątek, M.; Niżnikowski, R.; Szymańska, Ż.; Bednarczyk, M.; Orłowski, E.; Morales-Villavicencio, A.; Ślęzak, M. The effect of aging and muscle type on the quality characteristics and lipid oxidation of lamb meat. *Arch. Anim. Breed.* **2019**, *62*, 383–391. [CrossRef]
20. Pérez-Jiménez, J.; Arranz, S.; Tabernero, M.; Díaz-Rubio, M.E.; Serrano, J.; Goñi, I.; Saura-Calixto, F. Updated methodology to determine antioxidant capacity in plant foods, oils and beverages: Extraction, measurement and expression of results. *Food Res. Int.* **2008**, *41*, 274–285. [CrossRef]
21. Sañudo, C.; Muela, E.; Campo, M.M. Key factors involved in lamb quality from farm to fork in europe. *J. Integr. Agric.* **2013**, *12*, 1919–1930. [CrossRef]
22. European Union. *European Community Standards for the Classification of Light Lambs Carcasses, Brochure No. CM84-94-703-ES-D (L-2985)*; Publishing Bureau of the European Communities: Luxembourg, 1994.

23. Tarladgis, B.G.; Watts, B.M.; Younathan, M.T.; Dugan, L. A distillation method for the quantitative determination of malonaldehyde in rancid foods. *J. Am. Oil Chem. Soc.* **1960**, *37*, 44–48. [CrossRef]
24. Folch, J.; Lees, M.; Sloane Stanley, G.H. A simple method for the isolation and purification of lipids from animal tissues. *J. Biol. Chem.* **1957**, *226*, 497–509. [CrossRef]
25. Brand-Williams, W.; Cuvelier, M.E.; Berset, C. Use of a free radical method to evaluate antioxidant activity. *LWT Food Sci. Technol.* **1995**, *28*, 25–30. [CrossRef]
26. Singleton, V.L.; Rossi, J.A. Colorimetry of total phenolics with phosphomolybdic-phosphotungstic acid reagents. *Am. J. Enol. Vitic.* **1965**, *16*, 144–158.
27. European Union. *Regulation (EU) No 1151/2012 of the European Parliament and of the Council for the Name 'Cordero Manchego' (PGI) (C 242/5)*; Official Journal of the European Union: Brussels, Belgium, 2019.
28. Holman, B.W.B.; Kashani, A.; Malau-Aduli, A.E.O. Growth and body conformation responses of genetically divergent australian sheep to spirulina (*Arthrospira platensis*) supplementation. *Am. J. Exp. Agric.* **2012**, *2*, 160–173. [CrossRef]
29. Hopkins, D.L.; Clayton, E.H.; Lamb, T.A.; van de Ven, R.J.; Refshauge, G.; Kerr, M.J.; Bailes, K.; Lewandowski, P.; Ponnampalam, E.N. The impact of supplementing lambs with algae on growth, meat traits and oxidative status. *Meat Sci.* **2014**, *98*, 135–141. [CrossRef] [PubMed]
30. El-Sabagh, M.R.; Abd Eldaim, M.; Mahboub, H.D.H.; Abdel-Daim, M. Effects of *Spirulina platensis* algae on growth performance, antioxidative status and blood metabolites in fattening lambs. *J. Agric. Sci.* **2014**, *6*, 92–98. [CrossRef]
31. Jaborek, J.R.; Zerby, H.N.; Moeller, S.J.; Fluharty, F.L. Effect of energy source and level, and sex on growth, performance, and carcass characteristics of lambs. *Small Rumin. Res.* **2017**, *151*, 117–123. [CrossRef]
32. Peña, F.; Avilés, C.; Domenech, V.; González, A.; Martínez, A.; Molina, A. Effects of stress by unfamiliar sounds on carcass and meat traits in bulls from three continental beef cattle breeds at different ageing times. *Meat Sci.* **2014**, *98*, 718–725. [CrossRef]
33. Mordenti, A.L.; Brogna, N.; Canestrari, G.; Bonfante, E.; Eusebi, S.; Mammi, L.M.E.; Giaretta, E.; Formigoni, A. Effects of breed and different lipid dietary supplements on beef quality. *Anim. Sci. J.* **2019**, *90*, 619–627. [CrossRef] [PubMed]
34. Silva Sobrinho, A.G.d.; Purchas, R.W.; Kadim, I.T.; Yamamoto, S.M. Características de qualidade da carne de ovinos de diferentes genótipos e idades ao abate. *Rev. Bras. de Zootec.* **2005**, *34*, 1070–1078. [CrossRef]
35. Vergara, H.; Gallego, L. Effect of electrical stunning on meat quality of lamb. *Meat Sci.* **2000**, *56*, 345–349. [CrossRef]
36. Nieto, G.; Díaz, P.; Bañón, S.; Garrido, M.D. Dietary administration of ewe diets with a distillate from rosemary leaves (*Rosmarinus officinalis* L.): Influence on lamb meat quality. *Meat Sci.* **2010**, *84*, 23–29. [CrossRef] [PubMed]
37. Vergara, H.; Bórnez, R.; Linares, M.B. CO2 stunning procedure on Manchego light lambs: Effect on meat quality. *Meat Sci.* **2009**, *83*, 517–522. [CrossRef]
38. Blanco, C.; Bodas, R.; Prieto, N.; Andrés, S.; López, S.; Giráldez, F.J. Concentrate plus ground barley straw pellets can replace conventional feeding systems for light fattening lambs. *Small Rumin. Res.* **2014**, *116*, 137–143. [CrossRef]
39. Ripoll, G.; Joy, M.; Muñoz, F. Use of dietary vitamin E and selenium (Se) to increase the shelf life of modified atmosphere packaged light lamb meat. *Meat Sci.* **2011**, *87*, 88–93. [CrossRef]
40. Pearce, K.L.; Rosenvold, K.; Andersen, H.J.; Hopkins, D.L. Water distribution and mobility in meat during the conversion of muscle to meat and ageing and the impacts on fresh meat quality attributes-A review. *Meat Sci.* **2011**, *89*, 111–124. [CrossRef]
41. Kim, Y.H.B.; Warner, R.D.; Rosenvold, K. Influence of high pre-rigor temperature and fast pH fall on muscle proteins and meat quality: A review. *Anim. Prod. Sci.* **2014**, *54*, 375–395. [CrossRef]
42. Lawrie, R.A.; Ledward, D. *Lawrie's Meat Science*, 7th ed.; Lawrie, R.A., Ed.; Woodhead Publishing: Cambridge, UK, 2006. [CrossRef]
43. Shorthose, W.R.; Powell, V.H.; Harris, P.V. Influence of electrical-stimulation, cooling rates aging on the shear force values of chilled lamb. *J. Food Sci.* **1986**, *51*, 889–892. [CrossRef]
44. Linares, M.B.; Bórnez, R.; Vergara, H. Effect of stunning systems on meat quality of Manchego suckling lamb packed under modified atmospheres. *Meat Sci.* **2008**, *78*, 279–287. [CrossRef] [PubMed]
45. Bhat, Z.F.; Morton, J.D.; Mason, S.L.; Bekhit, A.E.-D.A. Role of calpain system in meat tenderness: A review. *Food Sci. Hum. Wellness* **2018**, *7*, 196–204. [CrossRef]
46. Tapp, W.N.; Yancey, J.W.S.; Apple, J.K. How is the instrumental color of meat measured? *Meat Sci.* **2011**, *89*, 1–5. [CrossRef]
47. Camo, J.; Beltrán, J.A.; Roncalés, P. Extension of the display life of lamb with an antioxidant active packaging. *Meat Sci.* **2008**, *80*, 1086–1091. [CrossRef]
48. Campo, M.M.; Nute, G.R.; Hughes, S.I.; Enser, M.; Wood, J.D.; Richardson, R.I. Flavour perception of oxidation in beef. *Meat Sci.* **2006**, *72*, 303–311. [CrossRef]
49. Muíño, I.; Apeleo, E.; de la Fuente, J.; Pérez-Santaescolástica, C.; Rivas-Cañedo, A.; Pérez, C.; Díaz, M.T.; Cañeque, V.; Lauzurica, S. Effect of dietary supplementation with red wine extract or vitamin E, in combination with linseed and fish oil, on lamb meat quality. *Meat Sci.* **2014**, *98*, 116–123. [CrossRef] [PubMed]
50. Luciano, G.; Vasta, V.; Monahan, F.J.; López-Andrés, P.; Biondi, L.; Lanza, M.; Priolo, A. Antioxidant status, colour stability and myoglobin resistance to oxidation of longissimus dorsi muscle from lambs fed a tannin-containing diet. *Food Chem.* **2011**, *124*, 1036–1042. [CrossRef]
51. Munir, N.; Sharif, N.; Naz, S.; Manzoor, F. Algae: A potent antioxidant source. *Sky J. Microbiol. Res.* **2013**, *1*, 22–31.

Article

Changes in Physical Meat Traits, Protein Solubility, and the Microstructure of Different Beef Muscles during Post-Mortem Aging

Yong-Hong Feng [†], Song-Shan Zhang [†], Bao-Zhong Sun, Peng Xie, Kai-Xin Wen and Chen-Chen Xu *

Institute of Animal Sciences, Chinese Academy of Agricultural Sciences, Beijing 100093, China; yike556@163.com (Y.-H.F.); zhangsongshan@caas.cn (S.-S.Z.); sunbaozhong@caas.cn (B.-Z.S.); seulbird@163.com (P.X.); wenkaixin215@126.com (K.-X.W.)
* Correspondence: chenxu30510@cau.edu.cn; Tel.: +86-010-62816010
† These authors contributed equally to this work.

Received: 19 April 2020; Accepted: 17 June 2020; Published: 19 June 2020

Abstract: This study was performed to compare the differences in pH, myofibril fragmentation index (MFI), total protein solubility (TPS), sarcoplasmic protein solubility (SPS), myofibrillar protein solubility (MPS), and the microstructure of seven beef muscles during aging. From the six beef carcasses of Xinjiang brown cattle, a total of 252 samples from *semitendinosus* (ST), *longissimus thoracis* (LT), *rhomboideus* (RH), *gastrocnemius* (GN), *infraspinatus* (IN), *psoas major* (PM), and *biceps femoris* (BF) muscles were collected, portioned, and assigned to six aging periods (1, 3, 7, 9, 11, and 14 day/s) and 42 samples were used per storage period. IN muscle showed the highest pH ($p < 0.05$) from 1 to 14 days and the lowest TPS ($p < 0.01$) from 9 to 14 days with respect to the other muscles. Moreover, the changes in IN were further supported by transmission electron microscopy due to the destruction of the myofibril structure. The highest value of MFI was tested in ST muscle from 7 to 14 days. The total protein solubility in PM, RH, and GN muscles were not affected ($p > 0.05$) as the aging period increased. The lowest TPS was found in the RH muscle on day 1, 3, and 7 and in the IN muscle on day 9, 11, and 14. The pH showed negative correlations with the MFI, TPS, and MPS ($p < 0.01$). The results suggest that changes in protein solubility and muscle fiber structure are related to muscle location in the carcass during aging. These results provide new insights to optimize the processing and storage of different beef muscles and enhance our understanding of the biological characteristics of Xinjiang brown cattle muscles.

Keywords: beef muscle; protein solubility; myofibril fragmentation; microstructure; aging

---

1. Introduction

Meat and meat products are important nutrient-intensive foods that are widely consumed worldwide [1]. For consumers, tenderness is a critical indicator that drives their willingness to repurchase and acceptability [2,3]. There are still great variations in the tenderness between different parts of beef muscles [4,5]. In the beef industry, aging is often considered to be one of the important factors determining the ultimate tenderness of the meat [6,7]. Due to the differences between the biochemical characteristics of these muscles, they may cause different degrees of response to aging [4]. Therefore, the inconsistency of the tenderness of different muscles has become the main problem that currently exists in the meat processing industry during the aging process.

Reducing the intake of fat in meat is essential for maintaining consumer health and reducing the risk of illness [8]. In fact, the type of muscle fiber is one of the factors that affects the lean meat rate [9]. The muscle fiber type differs greatly among various body parts and is generally divided into slow-oxidized fiber (type I), fast oxido-glycolytic fiber (type IIA), and fast glycolytic fibers (type IIB and type IIX) [10,11]. Studies indicated that muscle fiber type was closely associated with muscle quality and more type IIB fiber might lead to poorer muscle quality and lower lean meat productivity [12,13]. Our previous unpublished study has shown that *semitendinosus* (ST) and *longissimus thoracis* (LT) muscles of Xinjiang brown cattle have large percentages of type IIB fibers (approximately 47.89% and 40.85%, respectively), *rhomboideus* (RH) and *gastrocnemius* (GN) muscles have high percentages of type IIA fibers (approximately 61.43% and 42.10%, respectively), and *infraspinatus* (IN), *psoas major* (PM), and *biceps femoris* (BF) muscles have large percentages of type I fibers (79.69%, 43.06%, and 39.75%, respectively). In the current study, we focused on these seven muscles, representing the characteristics of the three muscle fiber types.

Protein in skeletal muscle is a key factor in determining the type of muscle fiber. Establishing the relationship between muscle protein structure and functional properties will contribute to the understanding of the modification mechanism of meat products during processing and storage [14]. According to the solubility and location of protein, it can be divided into myofibrillar protein, sarcoplasmic protein, and matrix protein, which maintain the structural integrity of myofibrillar fiber [15,16]. These proteins are susceptible to oxidative degradation during aging, which may explain the meat tenderization after muscle fiber breakage [17]. Oxidation affects the chemical properties of proteins and changes in protein solubility can reflect the degree of protein denaturation [18,19]. Sarcoplasmic protein solubility (SPS) is sometimes used as a measure of muscle quality [20]. It has been well documented that precipitated or denatured sarcoplasmic proteins may bind to myofibrils, leading to a decrease in water holding capacity [21–23]. It is vital to the production of Xinjiang brown beef, for the lives of Xinjiang people, and for local meat production. However, there are still limited studies examining the changes in protein solubility between various parts of muscles of Xinjiang brown cattle during post-mortem aging.

Therefore, the aim of this research was to evaluate the changes in pH, myofibril fragmentation index (MFI), total protein solubility (TPS), SPS and myofibrillar protein solubility (MPS), and the microstructure of seven beef muscles (ST, LT, RH, GN, IN, PM, BF) during aging from 1 to 14 days.

2. Materials and Methods

2.1. Animals and Muscle Samples Preparation

Six Xinjiang brown bull calves of approximately 30 months of age were slaughtered at a local slaughterhouse (Yining, Xinjiang, China) according to the commercial procedures. The mean weight at slaughter was 566 ± 32 kg. All corn-fed calves were raised on the same farm to ensure background consistency and then slaughtered using electrical stunning on the same day. The carcasses were overhung in a cold room (3 ± 1 °C) for 24 h. All procedures were undertaken following the guidelines given by the Animal Care and Ethics Committee for animal experiments, Institute of Animal Science, Chinese Academy of Agricultural Sciences.

The *semitendinosus* (ST), *longissimus thoracis* (LT), *rhomboideus* (RH), *gastrocnemius* (GN), *infraspinatus* (IN), *psoas major* (PM), and *biceps femoris* (BF) muscles were collected from each carcass. The samples were kept on ice and transported to the laboratory after 50 min and all visible intermuscular and subcutaneous fat was removed. Prior to packaging, each sample was stamped with a date mark. Each muscle was further separated into three equal-length sections, resulting in six muscle Sections (5 cm × 3 cm × 2 cm, 50 ± 0.05 g) per carcass, with 42 slices per aging period (six replications per muscle and per aging period). Afterwards, the samples were individually vacuum-packaged in a polyolefin bag and randomly assigned to aging at 3 ± 1 °C (relative humidity: 70–80%) for 1, 3, 7, 9, 11, and 14 days (within muscles, samples from the same carcass were not aged for the same time). Upon completion of each postmortem aging period, samples were taken to determine pH and transmission electron microscopy (TEM). The remaining samples were stored at −30 °C for further analysis within 2 weeks.

2.2. pH Measurement

The pH value of the beef muscles was measured at 1, 3, 7, 9, 11, and 14 days postmortem following the method described by Kim et al. [24], with minor modifications. Chopped meat (10 g) was mixed with 100 mL of distilled water for 15 s and then homogenized using an Ultra-Turrax T25 homogenizer (IKA-Werke, Gmbh & Co., Staufen in Breisgau, Germany) at 2800 g. The pH of the homogenate was measured using a pH meter equipped with an electrode (PB-10, the precision was 0.01, made by Sartorius Group, Goettingen, Germany). Each sample was determined three times.

2.3. Myofibril Fragmentation Index (MFI)

MFI was determined by modification of the method by Hou et al. [25]. Two gram samples were homogenized with a homogenizer (FJ200-S, Shanghai Specimen Model Co., Shanghai, China) at 10,000 g for 60 s (3 × 20 s with a 60 s break between bursts) at 4 ± 2 °C in 20 mL ice-cold buffer (100 mM KCl, 20 mM K_2HPO_4, 1 mM EGTA, 1 mM $MgCl_2$, and 1 mM NaN_3, pH 7.0). The homogenates were centrifuged using a LG10-24A model centrifuge (Peking Medical centrifuge Factory, Beijing, China) at 3000 g for 15 min at 4 °C and the supernatant was discarded. The pellets were homogenized in 20 mL of homogenizing buffer and centrifuged and the supernatant was discarded again. The resulting pellets were then resuspended in 5 mL of homogenizing buffer and filtered through a polyethylene strainer (200-mesh) to remove the fat and connective tissue. Then, 5 mL buffer was used to promote the passage of myofibrils through the strainer. The protein concentration of suspension was determined by the biuret method [26]. The protein concentration was diluted to 0.5 mg/mL and measured spectrophotometrically at 540 nm (UV 6100, Metash, Shanghai, China). MFI was calculated by multiplying A540 by 200.

2.4. Protein Solubility

The total protein solubility (TPS) and sarcoplasmic protein solubility (SPS) were measured using the procedure of Li et al. [27]. A 1 g sample of each muscle (six replications per muscle and per aging period) was homogenized with 10 mL cold 0.025 M potassium phosphate buffer (pH 7.2), which was then shaken at 4 °C for 12 h and centrifuged at 1500 g for 20 min. The total protein concentrations of the supernatants were assessed using the biuret method [26]. The step of SPS detection was that the muscle sample (1 g) was homogenized in ice-cold 20 mL 0.025 M potassium phosphate buffer (pH 7.2) and then carried out using the same shaking, homogenization, and centrifugation procedures mentioned above. Myofibrillar protein solubility (MPS) was calculated by TPS minus SPS. The value was expressed as micrograms of soluble protein per gram of meat.

2.5. Microstructure Analysis

At each storage time point, strips (approximately 5 × 1 × 1 mm) were cut off from three muscles (ST, RH, and IN) and were selected using transmission electron microscopy (TEM). Muscle samples were fixed with 2.5% glutaraldehyde in 0.1 M phosphate buffer (pH 7.2) and 2% osmium tetroxide. Then, the gradually increasing concentrations of ethanol (50–100%, *v/v*) were used for dehydration. The solvent was then switched to acetone and the samples were dehydrated with acetone gradient, followed by replacing with propylene oxide and being embedded with resin. The samples were trimmed, sectioned (50–70 nm), and stained with uranyl acetate and lead citrate. The samples were viewed under TEM (Hitachi, H-7500, Japan) at required magnification as per the standard procedures.

2.6. Statistical Analysis

The experimental design was a randomized complete block design, where each carcass ($n = 6$) served as a block. Each measurement was performed in triplicate. To determine the significant effect by two factors (muscles and aging period), data analysis was performed using the SAS 9.2 program (SAS Institute, Cary, NC, USA) with muscles and aging period as the main effects, using two-way analysis of variance (ANOVA). Analysis of variance was performed on all the variables using the General Linear Model (GLM) procedure. Duncan's multiple range test ($p < 0.05$) was used to determine the significance of the differences in the mean values for different samples. Pearson's correlation coefficients were calculated for variables.

3. Results and Discussions

3.1. Changes in pH Value

The changes in pH values of beef muscles during aging are shown in Table 1. There was a considerable variation in pH between muscles. Compared to the IN muscle, the pH values in ST, LT, PM, and BF were lower at 1 day ($p < 0.05$). The difference in pH may be due to the rate of glycolysis between muscles [28]. Similar results were obtained in some studies of beef and lamb [29,30], indicating that the rate of decrease in the pH of PM accelerated. Aging had no significant impact on pH in PM muscles ($p > 0.05$) from 1 to 14 days. Except for IN and PM, no significant changes were observed ($p > 0.05$) in the pH of the remaining six muscles from 3 to 14 days of aging. In terms of value, the pH of IN showed the highest value ($p < 0.05$) of all the muscles in the whole aging period. IN had more type I muscle fibers and its main characteristics were higher oxidative metabolism and lower glycogen content [31]. Therefore, in this study, we can speculate that lower glycolysis capacity would result in a higher pH of IN muscle, which might influence proteolysis.

Table 1. Changes in pH values of different beef muscles during aging.

Items	Aging (Days)						SEM
	1	3	7	9	11	14	
ST	5.65 [a,y]	5.45 [ab,x]	5.54 [b,xy]	5.50 [a,xy]	5.45 [ab,x]	5.38 [a,x]	0.028
LT	5.64 [a,y]	5.38 [a,x]	5.41 [a,x]	5.39 [a,x]	5.37 [a,x]	5.40 [ab,x]	0.027
RH	5.76 [ab,y]	5.55 [bc,x]	5.53 [b,x]	5.62 [ab,x]	5.56 [b,x]	5.55 [ab,x]	0.028
GN	5.72 [ab,y]	5.46 [ab,x]	5.53 [b,x]	5.51 [a,x]	5.38 [a,x]	5.45 [ab,x]	0.032
IN	5.87 [b,z]	5.66 [c,xy]	5.70 [c,xy]	5.76 [b,yz]	5.73 [c,xyz]	5.56 [b,x]	0.029
PM	5.59 [a]	5.51 [ab]	5.53 [b]	5.58 [ab]	5.51 [ab]	5.55 [ab]	0.022
BF	5.67 [a,y]	5.45 [ab,x]	5.42 [a,x]	5.47 [a,x]	5.43 [ab,x]	5.51 [ab,x]	0.026
SEM	0.026	0.023	0.022	0.038	0.030	0.022	

ST, *semitendinosus*; LT, *longissimus thoracis*; RH, *rhomboideus*; GN, *gastrocnemius*; IN, *infraspinatus*; PM, *psoas major*; BF, *biceps femoris*. [a–c] Values within a column with different superscript are significantly different ($p < 0.05$). [x–z] Values within a row with different superscript are significantly different ($p < 0.05$). SEM, standard error of the mean. $n = 6$ in each muscle at each aging period.

3.2. Changes in Myofibril Fragmentation Index (MFI)

Changes in MFI of different beef muscles during aging are displayed in Table 2. The highest MFI was observed in the RH muscle at 1 day ($p < 0.05$). The ST muscle had the highest MFI from day 7 onwards ($p < 0.05$). Compared with the other muscles, the MFI values of IN and LT were found to be lower from 3 to 11 days ($p < 0.05$). MFI is used as an important indicator of I band rupture and interstitial fibril connection breakage [32]. These results seem to indicate that the degree of rupture of muscle fibers in the I band is greater in ST muscles during aging, resulting in changes in the integrity and solubility of myofibrillar proteins [33,34]. The MFI of each individual muscle increased significantly with the aging period ($p < 0.05$) and this is in agreement with Li et al. [35]. However, the difference in MFI change was very small in the later days of the aging period. This might be due to the fact that as the aging period increases, the degree of variation in myofibril breaks becomes smaller [36].

Table 2. Changes in the myofibril fragmentation index (MFI) of different beef muscles during aging.

Items	Aging (days)						SEM
	1	3	7	9	11	14	
ST	102.57 [d,x]	107.93 [c,x]	154.13 [d,y]	157.53 [e,y]	156.50 [e,y]	189.93 [e,z]	7.47
LT	81.87 [bc,x]	83.10 [a,x]	86.97 [a,xy]	88.27 [a,xy]	88.90 [a,xy]	92.60 [a,y]	1.17
RH	122.23 [e,x]	129.93 [d,xy]	131.73 [c,xy]	132.97 [cd,xy]	133.77 [c,xy]	143.43 [cd,y]	2.11
GN	87.97 [c,x]	105.87 [c,y]	128.56 [c,z]	131.70 [c,z]	131.37 [c,z]	136.93 [c,z]	4.41
IN	68.80 [a,x]	76.87 [a,xy]	79.77 [a,xy]	84.37 [a,y]	87.63 [a,y]	101.83 [ab,z]	2.62
PM	75.90 [ab,x]	93.47 [b,y]	98.90 [b,y]	111.87 [b,z]	110.93 [b,z]	110.07 [b,z]	3.27
BF	99.43 [d,w]	122.87 [d,x]	130.67 [c,x]	141.80 [d,y]	144.33 [d,yz]	152.47 [d,z]	4.33
SEM	3.95	4.19	5.77	5.76	5.62	7.07	

ST, *semitendinosus*; LT, *longissimus thoracis*; RH, *rhomboideus*; GN, *gastrocnemius*; IN, *infraspinatus*; PM, *psoas major*; BF, *biceps femoris*. [a–c] Values within a column with different superscript are significantly different ($p < 0.05$). [x–z] Values within a row with different superscript are significantly different ($p < 0.05$). SEM, standard error of the mean. $n = 6$ in each muscle at each aging period.

3.3. Changes in Protein Solubility

The results of the changes in TPS, SPS, and MPS during aging are shown in Table 3. No significant changes ($p > 0.05$) were evidenced in TPS of RH, GN, and PM muscles and in MPS of IN and BF muscles during aging. In regard to SPS, there were no significant differences in SPS between different muscles at day 11 and 14 ($p > 0.05$). IN muscle exhibited a lower TPS at day 9, 11, and 14 than that at day 7 ($p < 0.05$). LT muscles had the highest TPS and MPS among all muscles at day 14 ($p < 0.05$). The SPS of RH was significantly lower than that of IN and GN at day 7 and 9 ($p < 0.05$). Protein solubility was an important indicator of protein properties, which was attributed to protein denaturation [37]. In this study, the results indicated that IN muscle had a high extent of protein denaturation after 9 days [19]. The solubility of myofibrillar protein increased with the aging period up to day 7 and then did not change significantly. This is related to the characteristics of myofibrillar proteins [38,39]. During aging, myofibrillar protein bonds were weakened and more protein hydrolysis was released, which resulted in higher MPS [40]. However, myofibrillar proteins are gradually unfolded as the aging period increases, exposing hydrophobic groups and resulting in almost no change in MPS [41].

Table 3. Changes in protein solubility of different beef muscles during aging.

Items	Aging (days)						SEM
	1	3	7	9	11	14	
Total Protein Solubility, mg/g							
ST	172.77 [ab,x]	190.83 [bc,yz]	207.59 [c,z]	183.39 [ab,xy]	192.81 [ab,yz]	193.16 [bc,yz]	3.19
LT	181.08 [bcd,x]	198.36 [c,xy]	193.45 [b,xy]	195.42 [ab,xy]	200.21 [b,y]	200.06 [c,y]	2.42
RH	169.00 [a]	174.24 [a]	176.05 [a]	183.80 [ab]	181.94 [ab]	185.91 [ab]	2.25
GN	189.47 [d]	187.76 [abc]	196.72 [bc]	198.46 [b]	185.01 [ab]	194.29 [bc]	2.01
IN	176.24 [abc,x]	185.24 [abc,x]	195.85 [bc,y]	178.47 [a,x]	177.82 [a,x]	179.02 [a,x]	1.95
PM	183.20 [cd]	191.27 [bc]	187.48 [ab]	184.91 [ab]	187.58 [ab]	189.43 [abc]	1.30
BF	179.64 [bc,x]	179.67 [ab,x]	208.62 [c,y]	183.42 [ab,x]	186.39 [ab,x]	187.92 [ab,x]	2.94
SEM	1.69	2.25	2.67	2.31	2.48	1.77	
Sarcoplasmic Protein Solubility, mg/g							
ST	59.71 [abc,xy]	57.20 [ab,xy]	60.66 [b,y]	49.57 [ab,x]	50.90 [xy]	50.53 [xy]	1.51
LT	68.31 [c,y]	65.49 [b,y]	50.83 [a,x]	48.75 [ab,x]	51.99 [x]	51.51 [x]	2.17
RH	51.35 [a,xy]	59.33 [ab,y]	48.77 [a,x]	42.00 [a,x]	46.14 [x]	44.02 [x]	1.70
GN	62.45 [bc,yz]	58.45 [ab,yz]	67.01 [b,z]	55.05 [b,xy]	48.67 [x]	55.15 [xy]	1.73
IN	55.69 [ab,yz]	52.04 [a,xy]	62.61 [b,z]	52.65 [b,xy]	45.06 [x]	46.51 [xy]	1.76
PM	51.93 [a,y]	59.08 [ab,z]	47.68 [a,x]	48.06 [ab,x]	47.51 [x]	44.08 [w]	1.21
BF	55.29 [ab,x]	60.24 [ab,xy]	70.66 [b,y]	48.20 [ab,x]	49.23 [x]	57.48 [xy]	2.31
SEM	1.55	1.19	2.14	1.15	1.02	1.68	

Table 3. *Cont.*

Items	Aging (days)						SEM
	1	**3**	**7**	**9**	**11**	**14**	
	Myofibrillar Protein Solubility, mg/g						
ST	113.07 [a,x]	133.63 [b,y]	146.93 [b,y]	133.81 [ab,y]	141.91 [y]	142.63 [abc,y]	3.28
LT	112.77 [a,x]	132.87 [b,y]	142.62 [ab,y]	146.67 [b,y]	148.22 [y]	148.55 [c,y]	3.59
RH	117.66 [ab,x]	114.90 [a,x]	127.28 [a,xy]	141.80 [ab,y]	135.80 [y]	141.89 [abc,y]	3.10
GN	127.01 [de,x]	129.31 [b,xy]	129.71 [a,xy]	143.41 [ab,y]	136.34 [xy]	139.14 [abc,xy]	2.15
IN	120.55 [bc]	133.20 [b]	133.23 [ab]	125.82 [a]	132.75	132.52 [ab]	1.91
PM	131.27 [e,x]	132.18 [b,x]	139.80 [ab,xy]	136.85 [ab,xy]	140.06 [xy]	145.34 [bc,y]	1.62
BF	124.35 [cd]	119.44 [ab]	137.96 [ab]	135.22 [ab]	137.16	130.44 [a]	2.55
SEM	1.57	2.09	2.17	2.37	2.33	1.86	

ST, *semitendinosus*; LT, *longissimus thoracis*; RH, *rhomboideus*; GN, *gastrocnemius*; IN, *infraspinatus*; PM, *psoas major*; BF, *biceps femoris*. [a–c] Values within a column with different superscript are significantly different ($p < 0.05$). [x–z] Values within a row with different superscript are significantly different ($p < 0.05$). SEM, standard error of the mean. $n = 6$ in each muscle at each aging period.

3.4. Pearson Correlations

Correlation coefficients among pH, MFI, TPS, SPS, and MPS of all muscles during aging are presented in Table 4. MFI, TPS, and MPS are negatively correlated with pH ($p < 0.01$). MFI is positively correlated with MPS ($p < 0.05$), whereas no correlations with TPS and SPS are observed ($p > 0.05$). The correlation coefficients of TPS with MPS are higher than those of SPS, indicating that TPS was affected to a larger extent by the denaturation of MPS than that of SPS. These results suggest that pH is a more important determinant for protein solubility than MFI in this study. Moreover, the correlation can further explain the results of protein solubility and confirm that the variation in protein solubility of different muscles may be due to pH changes.

Table 4. Pearson correlation coefficients for pH, MFI, and protein solubility.

	MFI	**TPS**	**SPS**	**MPS**
pH	−0.314 **	−0.419 **	−0.032	−0.377 **
MFI		0.086	−0.144	0.183 *
TPS			0.300 **	0.743 **
SPS				−0.416 **

MFI, myofibril fragmentation index; TPS, total protein solubility; SPS, sarcoplasmic protein solubility; MPS, myofibrillar protein solubility. * $p < 0.05$; ** $p < 0.01$. $n = 6$ in each muscle at each aging period.

3.5. Changes in Muscle Microstructure

The changes in the microstructure of muscle (ST, RH, and IN) by TEM are shown in Figure 1. In ST, RH, and IN muscles, the sarcoplasmic reticulum around the sarcomeres can be clearly distinguished and the myofibrils are tightly combined with the visible I-band and A-band and the Z-disk and M-line can be differentiated on day 1 and day 3 (Figure 1(a1,a2,b1,b2,c1,c2)). After day 7, the overall integrity of the myofibrils diminishes (Figure 1(a3,b3,c3)). The M-line located on the centre of the A-band looks vague and the Z-disk and I-band junctions are weakened after day 9. The Z-disk is distorted but undamaged, although some longitudinal splits are evident in RH muscles (Figure 1(b4,b5)). The worst myofibrillar structure is observed in IN muscle, in which the overlapping structure of thick and thin filaments is destroyed, the skeletal muscle structure is severely broken, and the Z-disk is distorted and weakened (Figure 1(c4,c5)). At day 14, IN muscle shows fractures in the Z-disk, as well as fragmentation at the junction of the I-band and Z-disk (Figure 1(c6)). Pan and Yeh [42] found that cracking of muscle fibers and shortening of sarcomere could largely reduce the tenderness of the meat. Aside from mechanical damage, muscle fiber structure was also affected by endogenous proteases during aging [43,44]. Moczkowska et al. [45] had reported that BF muscles were more sensitive to oxidation than LL muscles and the extent of this phenomenon depends on the type of muscle examined. In this study, TEM results showed that IN muscle had a longer sarcomere length and a greater degree of rupture in the muscle fiber structure, as expected because this muscle has a higher proportion of type I fiber (approximately 79.69%), which predicted that the aging rate would be faster. These results are in accordance with the MFI described above. On the other hand, previous research shows that the concentration of coenzyme Q10, carnosine, and taurine in IN muscle was higher than that in LT muscle [46], implying a higher concentration of functional bioactive compounds in oxidized muscle. These results suggest that oxidized muscle accelerates the improvement of meat tenderness.

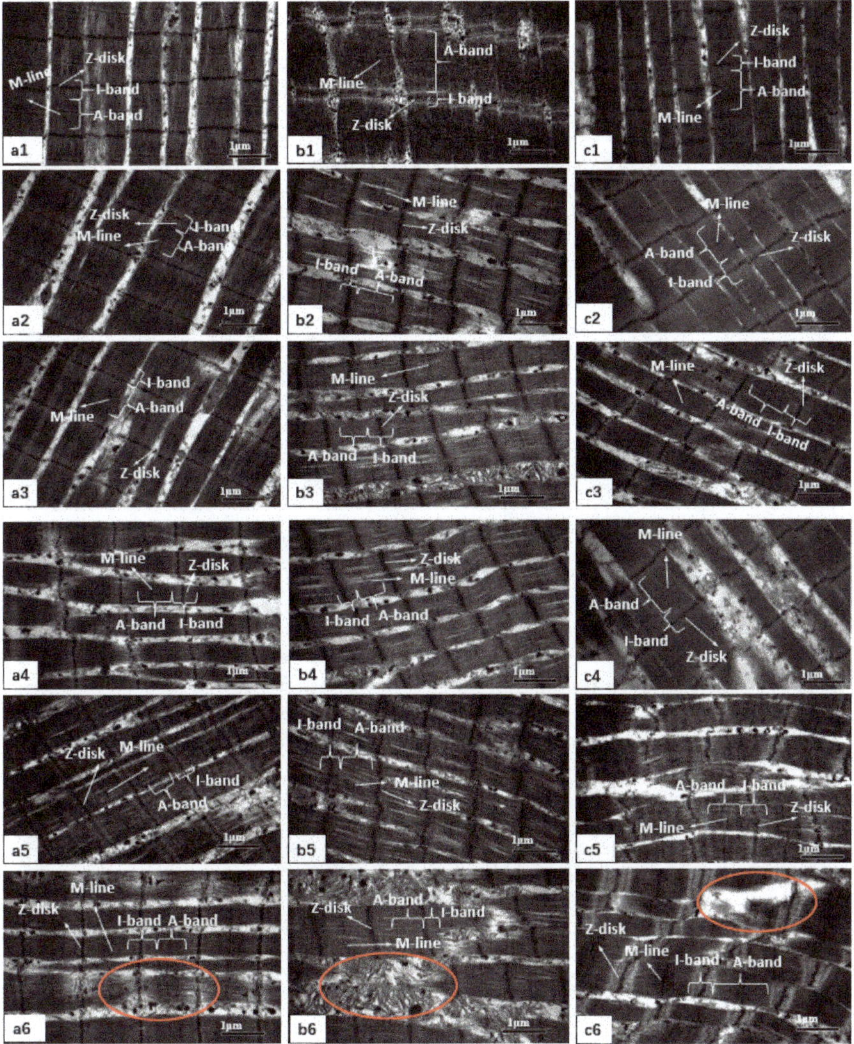

Figure 1. Transmission electron microscopy (TEM) images of beef muscles during different aging periods. TEM images at original magnifications of ×25,000. (**a1–a6**): The microstructure of *semitendinosus* (ST) at day 1, 3, 7, 9, 11, and 14. (**b1–b6**): The microstructure of *rhomboideus* (RH) at day 1, 3, 7, 9, 11, and 14. (**c1–c6**): The microstructure of *infraspinatus* (IN) at day 1, 3, 7, 9, 11, and 14. Red circles indicate the fracturing zones.

4. Conclusions

In this study, it can be concluded that the meat characteristics of different beef muscles responded differently to the storage time and were significantly affected by carcass position. The meat quality parameters of IN muscle were improved, which was due to its higher pH value, lower MFI and TPS, and greater degree of myofibril rupture, followed by ST and RH. However, there is little difference in meat quality parameters among other muscles. Therefore, for IN, it can be suggested that the characteristics are particularly suitable for producing high-quality beef products. Additionally, a significant interaction between pH and protein solubility was observed. Further research should be conducted to explore the mechanism of how pH affects protein solubility (for example, to assess its effect on cathepsin activity).

Author Contributions: Conceptualization, S.-S.Z.; methodology, Y.-H.F. and K.-X.W.; software, C.-C.X.; validation, C.-C.X.; formal analysis, C.-C.X. and S.-S.Z.; investigation, P.X.; resources, Y.-H.F.; data curation, Y.-H.F.; writing—original draft, Y.-H.F.; writing—review and editing, C.-C.X.; visualization, B.-Z.S.; supervision, S.-S.Z.; project administration, C.-C.X.; funding acquisition, P.X. All authors have read and agreed to the published version of the manuscript.

Funding: This study was funded by grants from the Program of National Beef Cattle and Yak Industrial Technology System (CARS-37) and the Modern Agricultural Industry Technology System and Beef Industry Innovation Construction Project of Hebei Province (HBCT2018130204).

Acknowledgments: The authors thank Haipeng Li for their valuable assistance during the muscle sampling work.

Conflicts of Interest: We declare that we have no conflict of interest.

References

1. Jiménez-Colmenero, F.; Herrero, A.; Cofrades, S.; Ruiz-Capillas, C. Meat: Eating Quality and Preservation. In *The Encyclopedia of Food and Health*; Caballero, B., Finglas, P., Toldrá, F., Eds.; Oxford Academic Press: Kidlington, UK, 2016; Volume 3, pp. 685–692.
2. Dobbs, L.M.; Jensen, K.L.; Leffew, M.B.; English, B.C.; Lambert, D.M.; Clark, C.D. Consumer willingness to pay for Tennessee beef. *J. Food Distrib. Res.* **2016**, *47*, 38–61.
3. O'Quinn, T.G.; Legako, J.F.; Brooks, J.C.; Miller, M.F. Evaluation of the contribution of tenderness, juiciness, and flavor to the overall consumer beef eating experience. *Trans. Anim. Sci.* **2018**, *2*, 26–36. [CrossRef]
4. Nair, M.N.; Canto, A.C.; Rentfrow, G.; Suman, S.P. Muscle-specific effect of aging on beef tenderness. *LWT Food Sci. Technol.* **2019**, *100*, 250–252. [CrossRef]
5. Von Seggern, D.D.; Calkins, C.R.; Johnson, D.D.; Brickler, J.E.; Gwartney, B.L. Muscle profiling: Characterizing the muscles of the beef chuck and round. *Meat Sci.* **2005**, *71*, 39–51. [CrossRef]
6. Neath, K.E.; Del Barrio, A.N.; Lapitan, R.M.; Herrera, J.R.V.; Cruz, L.C.; Fujihara, T.; Muroya, S.; Chikuni, K.; Hirabayashi, M.; Kanai, Y. Difference in tenderness and pH decline between water buffalo meat and beef during postmortem aging. *Meat Sci.* **2007**, *75*, 499–505. [CrossRef]
7. Hanzelková, Š.; Simeonovová, J.; Hampel, D.; Dufek, A.; Šubrt, J. The effect of breed, sex and aging time on tenderness of beef meat. *Acta Vet. Brno* **2011**, *80*, 191–196. [CrossRef]
8. Jiménez-Colmenero, F.; Herrero, A.M.; Ruiz-Capillas, C.; Cofrades, S. Meat and functional foods. In *Handbook of Meat and Meat Processing*; Hui, Y.H., Ed.; CRC Press, Taylor & Francis Group: Boca Raton, FL, USA, 2012; pp. 225–248.
9. Lee, S.H.; Joo, S.T.; Ryu, Y.C. Skeletal muscle fiber type and myofibrillar proteins in relation to meat quality. *Meat Sci.* **2010**, *86*, 166–170. [CrossRef]
10. Choi, Y.M.; Kim, B.C. Muscle fiber characteristics, myofibrillar protein isoforms, and meat quality. *Livest. Sci.* **2009**, *122*, 105–118. [CrossRef]
11. Jeong, J.Y.; Jeong, T.C.; Yang, H.S.; Kim, G.D. Multivariate analysis of muscle fiber characteristics, intramuscular fat content and fatty acid composition in porcine longissimus thoracis muscle. *Livest. Sci.* **2017**, *202*, 13–20. [CrossRef]
12. Hwang, Y.H.; Kim, G.D.; Jeong, J.Y.; Hur, S.J.; Joo, S.T. The relationship between muscle fiber characteristics and meat quality traits of highly marbled Hanwoo (Korean native cattle) steers. *Meat Sci.* **2010**, *86*, 456–461. [CrossRef]

13. Kim, G.D.; Yang, H.S.; Jeong, J.Y. Comparison of characteristics of myosin heavy chain-based fiber and meat quality among four bovine skeletal muscles. *Korean J. Food Sci. Anim. Resour.* **2016**, *36*, 819. [CrossRef] [PubMed]

14. Herrero, A.M.; Carmona, P.; Jiménez-Colmenero, F.; Ruiz-Capillas, C. Applications of vibrational spectroscopy to study protein structural changes in muscle and meat batter systems. In *Applications of Vibrational Spectroscopy to Food Science*; Chalmers, J., Griffiths, P., Li-Chan, E., Eds.; John Wiley & Sons: West Sussex, UK, 2010; pp. 315–328.

15. Smuder, A.J.; Kavazis, A.N.; Hudson, M.B.; Nelson, W.B.; Powers, S.K. Oxidation enhances myofibrillar protein degradation via calpain and caspase-3. *Free Radic. Biol. Med.* **2010**, *49*, 1152–1160. [CrossRef] [PubMed]

16. Veiseth-Kent, E.; Hollung, K.; Ofstad, R.; Aass, L.; Hildrum, K.I. Relationship between muscle microstructure, the calpain system, and shear force in bovine longissimus dorsi muscle. *J. Anim. Sci.* **2010**, *88*, 3445–3451. [CrossRef] [PubMed]

17. Hopkins, D.L.; Thompson, J.M. The degradation of myofibrillar proteins in beef and lamb using denaturing electrophoresis—An overview. *J. Muscle Foods* **2002**, *13*, 81–102. [CrossRef]

18. Marcos, B.; Kerry, J.P.; Mullen, A.M. High pressure induced changes on sarcoplasmic protein fraction and quality indicators. *Meat Sci.* **2010**, *85*, 115–120. [CrossRef]

19. Santos, M.D.; Delgadillo, I.; Saraiva, J.A. Extended preservation of raw beef and pork meat by hyperbaric storage at room temperature. *Int. J. Food Sci.* **2020**, *55*, 1171–1179. [CrossRef]

20. Joo, S.; Kauffman, R.; Kim, B.; Park, G. The relationship of sarcoplasmic and myofibrillar protein solubility to colour and water-holding capacity in porcine longissimus muscle. *Meat Sci.* **1999**, *52*, 291–297. [CrossRef]

21. Dai, Y.; Miao, J.; Yuan, S.Z.; Liu, Y.; Li, X.M.; Dai, R.T. Colour and sarcoplasmic protein evaluation of pork following water bath and ohmic cooking. *Meat Sci.* **2013**, *93*, 898–905. [CrossRef] [PubMed]

22. Liu, J.; Arner, A.; Puolanne, E.; Ertbjerg, P. On the water-holding of myofibrils: Effect of sarcoplasmic protein denaturation. *Meat Sci.* **2016**, *119*, 32–40. [CrossRef] [PubMed]

23. Wilson, G.G., III; van Laack, R.L.J.M. Sarcoplasmic proteins influence water-holding capacity of pork myofibrils. *J. Sci. Food Agric.* **1999**, *79*, 1939–1942. [CrossRef]

24. Kim, H.J.; Sujiwo, J.; Kim, H.J.; Jang, A. Effects of dipping chicken breast meat inoculated with Listeria monocytogenes in lyophilized scallion, garlic, and kiwi extracts on its physicochemical quality. *Food Sci. Anim. Resour.* **2019**, *39*, 418. [CrossRef] [PubMed]

25. Hou, X.; Liang, R.; Mao, Y.; Zhang, Y.; Niu, L.; Wang, R.; Liu, C.; Liu, Y.; Luo, X. Effect of suspension method and aging time on meat quality of Chinese fattened cattle M. *Longissimus Dorsi. Meat Sci.* **2014**, *96*, 640–645. [CrossRef]

26. Gornall, A.G.; Bardawill, C.J.; David, M.M. Determination of serum proteins by means of the biuret reaction. *J. Biol. Chem.* **1949**, *177*, 751–766. [PubMed]

27. Li, Y.J.; Li, J.L.; Zhang, L.; Gao, F.; Zhou, G.H. Effects of dietary starch types on growth performance, meat quality and myofibre type of finishing pigs. *Meat Sci.* **2017**, *131*, 60–67. [CrossRef]

28. Anderson, M.J.; Lonergan, S.M.; Fedler, C.A.; Prusa, K.J.; Binning, J.M.; Huff-Lonergan, E. Profile of biochemical traits influencing tenderness of muscles from the beef round. *Meat Sci.* **2012**, *91*, 247–254. [CrossRef]

29. Melody, J.L.; Lonergan, S.M.; Rowe, L.J.; Huiatt, T.W.; Mayes, M.S.; Huff-Lonergan, E. Early postmortem biochemical factors influence tenderness and water-holding capacity of three porcine muscles. *J. Anim. Sci.* **2004**, *82*, 1195–1205. [CrossRef] [PubMed]

30. Ilian, M.A.; Morton, J.D.; Kent, M.P.; Le Couteur, C.E.; Hickford, J.; Cowley, R.; Bickerstaffe, R. Intermuscular variation in tenderness: Association with the ubiquitous and muscle-specific calpains. *J. Anim. Sci.* **2001**, *79*, 122–132. [CrossRef]

31. Shen, L.Y.; Luo, J.; Lei, H.G.; Jiang, Y.Z.; Bai, L.; Li, M.Z.; Tang, G.Q.; Li, X.W.; Zhang, S.H.; Zhu, L. Effects of muscle fiber type on glycolytic potential and meat quality traits in different Tibetan pig muscles and their association with glycolysis-related gene expression. *Genet. Mol. Res.* **2015**, *14*, 14366–14378. [CrossRef]

32. McDonagh, M.B.; Fernandez, C.; Oddy, V.H. Hind-limb protein metabolism and calpain system activity influence post-mortem change in meat quality in lamb. *Meat Sci.* **1999**, *52*, 9–18. [CrossRef]

33. Knorr, D.; Zenker, M.; Heinz, V.; Lee, D.U. Applications and ultrasonics in food potential of processing. *Trends Food Sci. Technol.* **2004**, *15*, 261–266. [CrossRef]

34. Xiong, G.; Fu, X.; Pan, D.; Qi, J.; Xu, X.; Jiang, X. Influence of ultrasound-assisted sodium bicarbonate marination on the curing efficiency of chicken breast meat. *Ultrason. Sonochem.* **2020**, *60*, 104808. [CrossRef] [PubMed]

35. Li, K.; Zhang, Y.; Mao, Y.; Cornforth, D.; Dong, P.; Wang, R.; Zhu, H.; Luo, X. Effect of very fast chilling and aging time on ultra-structure and meat quality characteristics of Chinese Yellow cattle M. *Longissimus Lumborum*. *Meat Sci.* **2012**, *92*, 795–804. [CrossRef] [PubMed]

36. Aroeira, C.N.; Torres Filho, R.A.; Fontes, P.R.; Ramos, A.L.; Castillo, C.J.C.; Hopkins, D.L.; Ramos, E.M. Comparison of different methods for determining the extent of myofibrillar fragmentation of chilled and frozen/thawed beef across postmortem aging periods. *Meat Sci.* **2020**, *160*, 107955. [CrossRef] [PubMed]

37. Choi, Y.M.; Lee, S.H.; Choe, J.H.; Rhee, M.S.; Lee, S.K.; Joo, S.T.; Kim, B.C. Protein solubility is related to myosin isoforms, muscle fiber types, meat quality traits, and postmortem protein changes in porcine longissimus dorsi muscle. *Livest. Sci.* **2010**, *127*, 183–191. [CrossRef]

38. Li, F.; Wang, B.; Liu, Q.; Chen, Q.; Zhang, H.; Xia, X.; Kong, B. Changes in myofibrillar protein gel quality of porcine longissimus muscle induced by its stuctural modification under different thawing methods. *Meat Sci.* **2019**, *147*, 108–115. [CrossRef] [PubMed]

39. Qian, S.; Li, X.; Wang, H.; Mehmood, W.; Zhong, M.; Zhang, C.; Blecker, C. Effects of low voltage electrostatic field thawing on the changes in physicochemical properties of myofibrillar proteins of bovine *Longissimus dorsi* muscle. *J. Food Eng.* **2019**, *261*, 140–149. [CrossRef]

40. Li, S.; Zheng, Y.; Xu, P.; Zhu, X.; Zhou, C. L-Lysine and L-arginine inhibit myosin aggregation and interact with acidic amino acid residues of myosin: The role in increasing myosin solubility. *Food Chem.* **2018**, *242*, 22–28. [CrossRef]

41. Maity, I.; Rasale, D.B.; Das, A.K. Sonication induced peptide-appended bolaamphiphile hydrogels for in situ generation and catalytic activity of Pt nanoparticles. *Soft Matter* **2012**, *8*, 5301–5308. [CrossRef]

42. Pan, B.S.; Yeh, W.T. Biochemical and morphological changes in grass shrimp (Penaeus monodon) muscle following freezing by air blast and liquid nitrogen methods. *J. Food Biochem.* **1993**, *17*, 147–160. [CrossRef]

43. Zhang, M.; Li, F.; Diao, X.; Kong, B.; Xia, X. Moisture migration, microstructure damage and protein structure changes in porcine longissimus muscle as influenced by multiple freeze-thaw cycles. *Meat Sci.* **2017**, *133*, 10–18. [CrossRef]

44. Alahakoon, A.U.; Faridnia, F.; Bremer, P.J.; Silcock, P.; Oey, I. Pulsed electric fields effects on meat tissue quality and functionality. In *Handbook of Electroporation*; Miklavcic, D., Ed.; Springer: Cham, Switzerland, 2016; Volume 4, pp. 2455–2475.

45. Moczkowska, M.; Półtorak, A.; Montowska, M.; Pospiech, E.; Wierzbicka, A. The effect of the packaging system and storage time on myofibrillar protein degradation and oxidation process in relation to beef tenderness. *Meat Sci.* **2017**, *130*, 7–15. [CrossRef] [PubMed]

46. Purchas, R.W.; Zou, M. Composition and quality differences between the longissimus and infraspinatus muscles for several groups of pasture-finished cattle. *Meat Sci.* **2008**, *80*, 470–479. [CrossRef] [PubMed]

Article

Application of Fat-Tailed Sheep Tail and Backfat to Develop Novel Warthog Cabanossi with Distinct Sensory Attributes

Leo Nyikadzino Mahachi [1,2], Monlee Rudman [2], Elodie Arnaud [2,3,4], Voster Muchenje [1] and Louwrens Christiaan Hoffman [2,5,*]

[1] Department of Livestock and Pasture Science, University of Fort Hare, Private Bag X 1314, Alice 5700, South Africa; leonyikadzinomahachi@gmail.com or 23415177@sun.ac.za (L.N.M.); vmuchenje@ufh.ac.za (V.M.)
[2] Department of Animal Sciences, Stellenbosch University, Private Bag X1, Matieland, Stellenbosch 7602, South Africa; diehamsandwich@gmail.com (M.R.); elodie.arnaud@cirad.fr (E.A.)
[3] CIRAD, UMR QualiSud, Matieland, Stellenbosch 7602, South Africa
[4] Qualisud, Univ Montpellier, Avignon Université, CIRAD, Institut Agro, Université de La Réunion, 97715 Montpellier, France
[5] Centre for Nutrition and Food Sciences, Queensland Alliance for Agriculture and Food Innovation (QAAFI), Agricultural Mechanisation Building A 8115, Office 110, Gatton, QL 4343, Australia
* Correspondence: louwrens.hoffman@uq.edu.au; Tel.: +61-4-1798-4547

Received: 23 October 2020; Accepted: 25 November 2020; Published: 8 December 2020

Abstract: This study compared the use of pork backfat (PF) and fat-tailed sheep tail and backfat (SF) on the physicochemical, fatty acids and sensory attributes of warthog cabanossi. There were no differences between weight loss during drying, moisture content, pH, water activity, salt content and lipid oxidation between the cabanossi types. However, protein and ash contents were higher in PF cabanossi whilst fat content was higher in SF cabanossi. The PF cabanossi had higher polyunsaturated fatty acids (especially n-6), lower monounsaturated fatty acids whilst the saturated fatty acid content was similar between the two cabanossi products. The n-3:n-6 ratio was more beneficial in the SF cabanossi. The descriptive sensory analysis showed two distinct products where PF cabanossi scored higher for most attributes. Although SF cabanossi scored less for these attributes, this cabanossi had unique and acceptable sensory attributes. This study concluded that fat-tailed sheep tail and backfat could be used to produce a unique cabanossi product of acceptable quality.

Keywords: consumer acceptance; fatty acids; lipid oxidation; physicochemical attributes; sensory attributes; venison

---

1. Introduction

The increasing awareness of the deleterious health effects of fat and saturated fatty acids (SFA) composition of meat and meat products has contributed to advances in alteration of the fatty acid (FA) profile of these food commodities [1,2]. Modern consumers prefer lean meat with high levels of polyunsaturated fatty acids (PUFA), especially n-3 and n-6 FA. In the porcine industry, positive strides have been made to produce lean pigs that have a favourably higher content of PUFA in their meat. One of the methods employed to improve the FA profile of pork is appropriate feeding with feed containing favourable fatty acids [2–7].

Although high PUFA levels are beneficial, they are responsible for producing soft fat associated with a low melting point [8–10]. This negatively affects the processing properties, oxidative stability and shelf life of meat products [11–13]. Meat products containing soft fat consequently have reduced firmness, exhibit fat caps (fatting out) and have an oily/shiny appearance when packed [5,12,14] although different recipes, packaging and storage may also influence product quality [15,16]. Sensory attributes such as flavour and aroma are also negatively affected due to increased lipid oxidation of products developed using soft fat [12,13,17] especially in the case of salted/dried products such as cabanossi due to the pro-oxidant effect of salt [18]. Moreover, the processing of dry sausages manufactured with soft fat may be challenging as they may not achieve an adequate drying due to the liquefaction of fat which coats the lean particles [19,20]. Therefore, it is important to explore alternative fat sources which could potentially improve product quality.

Fat-tailed sheep reserve fat in their tails to be used during times when natural food resources are scarce [8,21]. However, management practices such as tail docking result in a shift of fat deposition site from the tail to muscle and subcutaneous tissues, and the latter increases backfat thickness [22–24]. Although fat derived from ruminants is predominantly SFA [2,4,9], feed restriction has been associated with improved fatty acid profile of sheep backfat [25,26]. Alves et al. [23] demonstrated that backfat from Damara (a fat-tailed sheep breed) under dietary restrictions, had 1.6% less palmitic acid (C16:0) mainly due to constrained de novo synthesis of lipids. On the other hand, the concentration of oleic acid (C18:1cis-9) increased due to stearoyl-CoA desaturase and lipogenic activity that occurs during backfat mobilization [25]. Moreover, van Harten et al. [24] observed higher levels of omega-3 FA, i.e., eicosapentaenoic acid (EPA; C20:5n−3), docosapentaenoic acid (DPA; 22:5n−3) and docosahexaenoic acid (DHA; C22:6n−3) in lipids from feed restricted Damara compared to non-fat-tailed sheep (Dorper and Australian Merino). Dietary manipulation is well known to improve the fatty acid profiles of sheep lipids [2,4,27]. Overall, lipids obtained from either fat-tailed sheep [8,26] or sheep fed PUFA-manipulated diets [2–4] have healthier FA profiles and could be valuable resources for developing meat products.

In various countries where the Muslim faith is the predominant faith and religious practices do not permit utility of pork in processed Halal foods, meat and fatty tissues from sheep and goats, including tail fat from fat-tailed sheep breeds are used to produce traditional and modern meat products [28–30]. In Iran, fat from fat-tailed sheep is used for cooking [8], whereas it is used to produce a variety of meat products in North Africa and Mediterranean regions [28]. Sheep fat has previously been used to produce beef-fermented sausages (Bez sucuk) [28,30] and is typically used to make droëwors (a typical South African dried sausage) [31] although its use in game meat products is still limited. While local people consider game meat as indigenous, it is also considered exotic and is attractive to adventurous consumers, particularly tourists, who want to experience new culinary experiences and take home products as souvenirs [32,33].

In South Africa, consumers enjoy indigenous meat products including droëwors, biltong (dried meat) and exotic meat products including salami (semi-dry fermented sausage) and cabanossi [33,34]. In that regard, some local artisanal manufacturers are acquainted with producing salami and cabanossi using game meat [35,36]. Cabanossi, alternatively kabanosy [33,37], is a semi-dry, cured and smoked pork sausage originating from Poland but has become a well-known snack among the South African population [33,36]. As of 2012, kabanosy was recognised by the European Union as a "Traditional Specialities Guaranteed" (TSG) product under the Commission Implementing Regulation No: 1044/2011 [38]. This registration, however, does not limit other producers from manufacturing the product although they may not use the same name if the recipe deviates from the original [36]. Traditionally, cabanossi was produced from a young fat pig (kabenek) without the addition of fat as opposed to the current practice of adding different fat sources to improve sensory attributes [36,39]. Nowadays, cabanossi is produced from several meat sources with additional fat for improved sensory attributes. Cabanossi produced from warthog meat and pork backfat (PF) was reported to be acceptable although criticized for its unhealthy FA profile [40], despite the favourable FA profile recorded for warthog meat [40,41], thereby stimulating research on other healthier fat sources. The aim of the current investigation was to determine the influence of fat-tailed sheep tail- and backfat (SF) on the physicochemical, fatty acids, lipid oxidation and sensory properties of warthog cabanossi.

2. Materials and Methods

2.1. Harvesting and Meat Sampling

Prior to the commencement of this study, ethical clearance was sought from the University of Fort Hare Research Ethics Committees (Ethical Clearance number: MUC361SMAH01). Warthogs (n = 24) used for meat in this study were harvested at a game farm (27° 22′ 09.26″ S, 31° 50′ 42.16″ E) near Pongola, KwaZulu Natal, South Africa. The farm falls within the Savannah biome, which is characterised by summer rainfall, mean annual precipitation ranging from 500–900 mm. The procedures for harvesting warthogs were described previously by Rudman et al. [42] Meat from entire warthog carcasses was used for cabanossi production except for the *m. longissimus thoracis et lumborum* muscle which was used by Rudman et al. [42] The SF used was sourced from pasture-fed Damara sheep while different batches of PF were purchased from a local abattoir in the Western Cape Province of South Africa. After harvesting and slaughter, meat and fat were vacuum packaged and stored at −20 °C until use.

2.2. Preparation of Cabanossi

The preparation of batter used to make cabanossi was similar to the method outlined previously [36]. Briefly, frozen warthog meat, PF and SF were thawed at <4 °C for ~12 h before use. Twelve separate 3 kg batches of PF (20%) and twelve separate 3 kg batches of SF (20%) cabanossi were produced. To avoid pseudoreplication, each batch of cabanossi was produced using meat from a unique animal (n = 24). To control the ratio between meat and fat in the batter, meat was trimmed of visible excess fat and sinews. Meat and fat were cut into ~5 × 5 cm cubes. Meat (2.4 kg) was weighed into a separate tray where 0.6 kg fat was added and hand mixed. After mixing, the mixture was minced through a 5 mm grinding plate (Manica, Model number CM-21, Equipamientos Carnicos, Barcelona, Spain) where after the spice mixture was added. The spice mixture consisted of 2% salt, 0.24% curing agent (Prague powder #1: Freddy Hirsh, Somerset West, South Africa), 0.2% black pepper, 0.06% nutmeg, 0.1% roasted and grounded caraway seeds and 0.2% mustard. The batter was mixed and minced again through the 5 mm plate before stuffing into 22 mm diameter sheep casings using a manual sausage filler (MOD. 7/V, Tre Spade, Torino, Italy). The cabanossi were hung in a temperature and humidity controlled drying chamber (Reich Airmaster® UKF 2000 BE, Reich Klima-Räuchertechnik, Urbach, Germany) for 16 h as described previously [33].

2.3. Physicochemical Analyses

The weight loss of each cabanossi batch was calculated as the percentage of lost weight relative to the initial weight of that batch. The pH of the raw batter was measured using a Crison 25 pH meter (Crison Instruments S.A., Alella, Spain) with an electrode probe. To determine the pH of the finished product, 3 g of sample was homogenised in 27 g dH_2O in duplicate before pH was measured using a Crison 25 pH meter with an electrode probe. Water activity was measured in duplicate using an Aqua Lab Due Point and Water Activity Meter 4TE (Decagon Devices, Inc., Pullman, WA, USA) at 25 °C. Moisture and ash were analysed according to the procedures of AOAC [43]. Fat was extracted using a chloroform/methanol (2:1 *v/v*) solution and determined according to Lee et al. [44] The defatted and dried meat samples (dried for at least 48 h at 60 °C in an oven) were analysed for nitrogen [43] using a calibrated LECO Nitrogen/Protein analyser (FP-528, Leco Corporation, St. Joseph, MI, USA). Protein was then calculated by multiplying the percentage of nitrogen by 6.25. All proximate analyses were performed per batch, in duplicate. Salt content was determined in duplicate by analysing the chloride concentration of each sample using a chloride analyser (Model 926, Sherwood Scientific, Cambridge, UK) after extraction from 0.3 g of sample in 50 mL 0.3 M nitric acid for at least 2 h.

2.4. Fatty Acid Composition

Fatty acid composition was determined by gas chromatography after extraction of lipids in chloroform/methanol (2:1 *v/v*) and transmethylation in methanol/sulphuric acid (19:1 *v/v*) as described by Neethling et al. [45] Fatty acid composition was expressed as a percentage of the content in the sample of total fatty acids. To assess the nutritional properties of the cabanossi, the ratios PUFA:SFA and n-6:n-3 were calculated. Lipid health indices (atherogenicity; AI and thrombogenicity; TI) of Ulbricht and Southgate [46] were also determined.

2.5. Lipid Oxidation

Lipid oxidation was determined by measuring thiobarbituric acid reactive substances (TBARS) on the raw batter and the finished cabanossi product using an acid-precipitation method previously described by members in our research group [47]. Absorbance was measured at 530 nm (Spectrostar Nano, BMG Labtech, Ortenberg, Germany) and TBARS were expressed as malondialdehyde (MDA) equivalent content (mg/kg).

2.6. Descriptive Sensory Analysis

Twelve replicates (n = 12) of each of the two cabanossi treatments were subjected to descriptive sensory analysis (DSA) by a panel of 12 judges trained according to approved procedures [48]. During training, the panellists made use of specific reference samples to formulate a list of sensory attributes in the order of aroma, flavour, appearance and texture (Table 1). The attributes for the descriptors were scored on an unstructured scale from 0–100. To cleanse the pallet between samples, panellists were given fresh apple quarters, water biscuits and spring water stored at room temperature (21 °C). After training, blind testing of the products was done over 12 replicate sessions that lasted six days. The judges sat in individual booths in a temperature (21 °C) and artificial daylight-controlled room. Each booth had a computer on which Compusense® five software (Compusense, Guelph, Canada) was installed.

2.7. Consumer Preference

The cabanossi were evaluated for preference to taste and appearance by 131 untrained consumers following a methodology described by Mahachi et al. [36] Each consumer was given two samples to evaluate; one from each treatment in the company of a corresponding questionnaire to fill and rate their preference for each sample. The questionnaire asked for demographic information and provided an unstructured scoring scale from 1–9 on which they could rate their preference for the different cabanossi treatments. All consumer identifiers were scrubbed from the data before analyses. The demographic information of the sample population is shown in Table 2.

2.8. Statistical Analysis

Physicochemical analyses data were statistically analysed using the generalised linear model procedures of SAS software (Version 9.4; SAS Institute Inc., Cary, NC, USA) in a completely randomised block design with the fat type as the main effect. Observations over time were combined in a split-plot analysis of variance with production stage (raw batter and finished product) as a sub-plot factor. For DSA data, judges (n = 12) were considered as block replicates for each sample (backfat type × replicate). A Shapiro–Wilk test was performed on the standardised residuals from the model to test for normality. In cases where there was significant deviation from normality, outliers were removed when the standardised residual for an observation deviated with more than three standard deviations from the model value. The data for consumer acceptance were analysed using mixed model repeated measures of ANOVA. Fisher's least significant difference was calculated at the 5% significance level to compare treatment means.

Table 1. Sensory attributes description, scale and reference standards used for the descriptive sensory analysis of two cabanossi treatments.

AROMA 0 = None, 100 = Prominent	Description of Attributes	Description of References
Overall aroma intensity	The overall intensity of aroma upon removing the cover	
Smokey aroma	Aroma associated with wood smoke	Low, moderate and high intensity liquid smoke solution [a]
Charred aroma	Aroma associated with blackened sheep meat cooked over high heat/flames	Braaied (barbequed) sheep meat
Fermented sour aroma	Aroma associated with all fermented meat products	Black forest ham/salami sticks
Cured pork aroma	Aroma associated with cured pork products	Cured unsmoked pork product (Ham)
Pork fat aroma	Aroma associated with roasted pork fat	Fat from roasted pork loin [b]
Mutton aroma	Aroma associated with mutton	Mutton chops [b]
Sheep-like fatty aroma	Aroma associated with mutton fat	Mutton fat [b]
Peppery aroma	Aroma associated with black pepper	Crushed black peppercorns
Sweet aroma	Sweet aroma associated with molasses	Low, moderate and high intensity molasses solution
Herbaceous aroma	General mixed herbs aroma	Rosemary, thyme, parsley and organum
FLAVOUR 0 = None, 100 = Prominent	**Description of Attributes**	**Description of References**
Overall flavour intensity	The overall intensity of flavour upon chewing	
Smokey flavour	Flavour associated with wood smoked pork	Fried bacon [c]
Woody flavour	Flavour associated with the wood from smoking	Wood sawdust solution [d]
Cured pork flavour	Flavour associated with cured pork products	Black forest ham/Fried bacon
Fermented sour flavour	Flavour associated with all fermented meat products	Black forest ham/Salami sticks
Pork fat flavour	Flavour associated with roasted pork fat	Roasted pork loin fat
Mutton flavour	Flavour associated with mutton	Mutton chops
Sheep-like fatty flavour	Flavour associated with mutton fat	Mutton fat
Peppery flavour	Flavour associated with black pepper	Crushed black peppercorns
Herbaceous flavour	General mixed herbs flavour	Rosemary, thyme, parsley and organum
Salty taste	Salty taste	Low, moderate, high intensity salt solution [e]
Sweet taste	Sweet aroma associated with molasses	Salami sticks
Fatty mouthfeel 0 = low, 100 = extremely high	The amount of fatty coating left on palate after swallowing	
APPEARANCE	**Description of Attributes**	**Description of References**
Red/brown colour intensity 0 = light, 100 = dark	Red/brown colour associated with different meat products	Dry cabanossi [f]
Fatty/Oily/Shininess 0 = dull, 100 = shiny	A measure of how oily the product looks (shininess)	Dry cabanossi/salami sticks
Percentage fat 0 = lean, 100 = fatty/abundant	The percentage perceived fat in the products	Dry cabanossi/pepper salami
TEXTURE	**Description of Attributes**	**Description of References**
First bite 0 = low, 100 = high	The amount of pressure required to bite through the cabanossi	Vienna's/salami
Sustained juiciness 0 = dry, 100 = extremely juicy	The impression of juiciness after first 5 chews	
Chewiness 0 = soft, 100 = extremely hard	The ease of chewing	Vienna's/salami sticks
Residue 0 = none, 100 = abundant	The amount of residue left in mouth after 10 chews	

[a] Low, moderate and high intensity were made using 1, 2 and 3 drops of liquid smoke solution in 100 mL water; [b] Roasted pork/mutton loin was roasted to an internal temperature of 75 °C and surrounding fat was used; [c] Fried bacon, pan fried bacon; [d] Wood sawdust solution, a cup-full (250 mL) of pine wood saw dust was soaked in water overnight and filtered; [e] Low, moderate, high intensity salt solution was made using 0.2%, 0.5% and 1% salt solution; [f] Dry cabanossi, a batch of cabanossi produced with no additional fat.

Table 2. Demographic variation of the sample population according to gender and age group.

Gender	Proportion *	Age Group	Proportion *
Male	60	18–23	20
Female	40	24–29	28
		30–39	22
		40–49	18
		50+	12
Total	100		100

* Proportion expressed as a percentage of the total population size.

3. Results and Discussion

3.1. Physicochemical Attributes

Results for the physicochemical analyses are presented in Table 3. There were no significant differences in weight loss during drying between treatments while moisture loss was higher ($p \leq 0.05$) for the PF treatment compared to the SF treatment. Moisture content was higher ($p \leq 0.01$) in the raw batter of the PF treatment compared to the SF treatment. This phenomenon may be explained by the fact that there were differences ($p \leq 0.01$) in the chemically analysed fat content of the two treatments' raw batter. Although similar amounts of fat were added during the mixing of batter, the sheep fat raw batter had more analysed fat compared to the pork fat treatment ($p \leq 0.01$). The suggestion would be that sheep subcutaneous and tail adipose tissue used in this study had more lipids and thus less moisture per unit of weight compared to that of pork. This could be attributed to de novo synthesis of lipids and fatty acids in sheep adipose tissue. Ruminant diets have a low fat content and hence most of their lipids and fatty acids are synthesised de novo in adipose tissue [22,49]. As a result, cabanossi end products showed a similar moisture content ($p > 0.05$). The lower moisture loss percentage of SF cabanossi could be related to saturation of sheep fat [50] or the fact that the fat content of PF raw batter was lower than that of the SF raw batter. Fat reduces water losses during drying of meat products by forming an oily coating around meat particles henceforth acting as "insulation" [36], which consequently limits the diffusion of moisture from the inside-out of the sausage [37,50].

Table 3. Physicochemical attributes and lipid oxidation of raw batter and cabanossi made with either pork backfat (PF) or sheep tail/backfat (SF).

Treatment	Raw Batter		*p*-Value	Finished Cabanossi Product		*p*-Value
	PF	SF		PF	SF	
Weight (kg)	2.6 ± 0.14	2.8 ± 0.16	0.302	1.7 ± 0.14	1.8 ± 0.21	0.097
Weight loss (%)	-	-	-	35.9 ± 4.75	33.2 ± 5.16	0.067
Moisture loss (%)	-	-	-	33.0 ± 0.85	29.9 ± 0.67	0.031
Water activity	-	-	-	0.94 ± 0.01	0.94 ± 0.01	0.146
pH	5.58 ± 0.17	5.56 ± 0.14	0.563	5.16 ± 0.10	5.14 ± 0.11	0.578
Moisture (%)	63.1 ± 0.88	60.7 ± 1.39	<0.0001	46.0 ± 1.16	46.2 ± 2.16	0.364
Protein (%)	19.9 ± 3.80	16.7 ± 3.34	0.030	27.8 ± 3.70	25.9 ± 2.09	0.032
Fat (%)	16.2 ± 4.28	20.5 ± 3.50	0.001	23.2 ± 4.40	24.8 ± 2.91	0.027
Ash (%)	2.7 ± 0.12	2.7 ± 0.10	0.502	4.0 ± 0.26	3.8 ± 0.22	0.006
Salt (%)	1.9 ± 0.11	1.9 ± 0.10	0.688	2.8 ± 0.25	2.7 ± 0.14	0.744
TBARS (mg/kg)	0.21 ± 0.07	0.24 ± 0.06	0.259	0.35 ± 0.09	0.37 ± 0.08	0.390

All data expressed as mean ± SE (n = 12).

Moisture content was within the range (39–50.7%) reported for commercial cabanossi [37,51], but lower than that reported by other authors for warthog and pork cabanossi (59% and 54%, respectively) [33]. Results from this study were also comparable to those reported in our companion paper (45.6%) [36] for warthog cabanossi produced with 20% PF, a similar amount as that used in the current investigation. Cabanossi were dried under temperatures that are not lethal to microorganisms, thus, like other meat products falling into this category, it should rely on the collective effects of safety hurdles of reduced water activity, pH and curing salts to prevent microbial spoilage [52,53]. These hurdles must be achieved in the order of: water activity <0.91 or pH < 4.5 or a combination of water activity <0.95 and pH < 5.2 [54]. These hurdles were achieved. Water activity was reduced in the final products but did not differ ($p > 0.05$) between treatments. Apart from the water activity we reported previously [36], which is comparable to the current study, no other literature was found reporting the water activity of cabanossi. This could be due to limited studies on cabanossi reported in the English language since this is originally a Polish product. Regarding pH, no differences were observed for raw batter and the finished product between treatments. Nevertheless, pH declined significantly ($p \leq 0.001$) after smoking and drying. The pH values obtained in this study are slightly higher than those observed for fermented sausages [55,56] where it is reported to fall below 5.0. This is expected because cabanossi is not fermented, i.e., no starter culture and/or sugars are added. In fermented sausages, pH declines as a function of increasing organic acids produced by predominantly lactic acid bacteria population growth from the starter cultures [57,58]. The pH of fermented meat sausages is expected to decline during drying and this allows it to reduce the rate of microbial spoilage [58]. Although not measured, the pH decline observed in this study could be attributed to the organic acid compounds of smoke onto the product during smoking [54].

Ash content in PF cabanossi was higher than in SF cabanossi while they were similar in the raw batter, which could be attributed to higher weight loss (although not significant) and moisture loss ($p \leq 0.01$). Additionally, the higher levels of total ash of the cabanossi end products (when compared to raw meat) are expected because the addition of salt and spices to raw batter increases the ash content [59]. Protein content was higher ($p \leq 0.05$) in PF cabanossi than sheep cabanossi as it was in the raw batter ($p \leq 0.05$). This is due to the fat:protein ratio being lower in low fat sausages as opposed to high fat sausages as previously reported [60].

3.2. Fatty Acid Composition

Table 4 shows results for the fatty acid composition of PF and SF cabanossi. The two most abundant SFA in the cabanossi products were palmitic acid (~22%) and stearic acid (~14%) and these fatty acids were similar in concentrations for both products. Similarly, the percentage total SFA did not differ between treatments. Total MUFA was higher and total PUFA lower in SF cabanossi compared to PF cabanossi ($p < 0.0001$) whereas the PUFA:SFA ratio was lower in the sheep cabanossi ($p < 0.0001$). The PUFA:SFA ratio is an important indicator of the healthiness of meat products and dietary guidelines suggest a ratio of no less than 0.4 [61]. Results from this study revealed that PF cabanossi could be beneficial in this regard. Linoleic acid was the most abundant PUFA in both treatments, though it was higher in PF cabanossi compared to the SF treatment ($p < 0.0001$). The abundance of linoleic acid in the pork backfat treatment is attributed to its occurrence in pork backfat [6,7,62], and warthog meat [40,41] where it is reported to be the most abundant PUFA.

Omega-6 fatty acids were notably higher in the PF cabanossi and there were no differences in the n-3 fatty acids between treatments whereas the n-6:n-3 ratio was lower ($p < 0.0001$) in the SF cabanossi. The n-6:n-3 ratio is also thought to be a good indicator of meat healthfulness [2,63]. It is recommended that healthy meat products should exhibit a ratio of less than 4.0 [64], therefore the SF cabanossi (2.65) produced in this study could be beneficial. A high n-6:n-3 ratio is linked to pathogenesis of some illnesses including certain cancers and some inflammatory and cardiovascular diseases while a lower ratio reduces the incidence of these ailments [65,66]. However, with regards to cabanossi, this may be less important because it is mostly consumed from time to time in limited quantities as a snack rather than regularly as food [36]. Other important indicators of the healthfulness of meat are the atherogenic index (AI) and thrombogenic index (TI) of Ulbricht and Southgate [44], which are determined on the basis that different fatty acids metabolise differently, either preventing or promoting atherosclerosis and coronary thrombosis [46,67]. Although the AI was similar between the two treatments, the TI was lower in the PF cabanossi ($p < 0.04$). Results from this study present an opportunity to label both warthog cabanossi products as healthier than several other dry-cured beef and pork meat products whose AI and TI range between 0.50–0.67 and 1.09–1.45, respectively [65,67].

Table 4. Percentage fatty acid composition of cabanossi made with either pork backfat (PF) or sheep tail/backfat (SF) before and after drying.

* FATTY ACID	Raw Batter		*p*-Value	Finished Cabanossi Product		*p*-Value
	PF	SF		PF	SF	
Saturated fatty acids						
C14:0 Myristic	1.69 ± 0.19	2.42 ± 0.48	0.001	1.38 ± 0.31	1.37 ± 0.30	0.356
C16:0 Palmitic	22.8 ± 1.34	21.95 ± 2.24	0.062	21.90 ± 1.92	22.16 ± 1.38	0.902
C18:0 Stearic	14.56 ± 2.21	12.79 ± 3.02	0.001	14.33 ± 2.03	14.27 ± 2.04	0.908
Monounsaturated fatty acids						
C16:1 Palmitoleic	2.10 ± 0.37	3.96 ± 1.08	0.000	1.9 ± 0.21	2.0 ± 0.43	0.741
C18:1n9c Oleic	23.49 ± 2.16	32.57 ± 5.10	<0.0001	24.4 ± 2.99	33.8 ± 2.75	0.637
Polyunsaturated fatty acids						
C18:2n6c Linoleic	22.32 ± 1.91	8.27 ± 6.15	<0.0001	21.47 ± 5.60	5.24 ± 0.65	<0.0001
C18:3n3 γ-α-Linolenic	2.59 ± 0.61	2.24 ± 0.49	0.030	2.58 ± 0.63	2.40 ± 0.77	0.040
SFA	42.95 ± 3.35	42.30 ± 4.53	0.864	43.18 ± 4.57	43.73 ± 5.80	0.580
MUFA	28.38 ± 2.50	43.30 ± 8.57	<0.0001	29.27 ± 3.84	45.38 ± 5.68	<0.0001
PUFA	28.67 ± 1.88	13.90 ± 6.15	<0.0001	27.55 ± 6.00	10.88 ± 1.32	<0.0001
PUFA: SFA	0.67 ± 0.09	0.32 ± 0.13	<0.0001	0.65 ± 0.18	0.25 ± 0.05	<0.0001
Total n-6	25.35 ± 1.85	10.96 ± 6.12	<0.0001	24.28 ± 5.78	7.82 ± 0.74	<0.0001
Total n-3	3.32 ± 0.64	2.94 ± 0.51	0.033	3.27 ± 0.71	3.07 ± 0.79	0.361
n-6:n-3	7.88 ± 1.49	3.79 ± 2.11	0.0001	7.33 ± 2.25	2.65 ± 0.50	<0.0001
Health indices						
Atherogenic index	0.43 ± 0.02	0.59 ± 0.07	<0.0001	0.44 ± 0.06	0.62 ± 0.04	0.050
Thrombogenic index	0.87 ± 0.12	0.94 ± 0.18	0.118	0.90 ± 0.19	0.99 ± 0.23	0.040

* Fatty acids: individual fatty acids with a percentage composition less than 1 were not displayed on the table but were included in calculation of total SFA, MUFA, PUFA, n-6 and n-3; All data are expressed as mean ± SE (n = 12).

3.3. Lipid Oxidation

Results for lipid oxidation are shown in Table 3. No differences were observed in terms of the TBARS detected in the raw batter and cabanossi products. The TBARS of cabanossi reported in this study (0.35 and 0.37 mg MDA equivalent/kg) are lower than those reported previously [59] for ostrich droëwors (7.99 mg MDA equivalent/kg) where pork backfat was used while Mukumbo et al. [47,68] reported values in pork droëwors reaching 0.7–3.8 mg MDA equivalent/kg dry matter (0.6–2.9 mg MDA equivalent/kg) at the end of drying. Deriving from these results, the extent of lipid oxidation during the manufacture of warthog cabanossi using these fat sources can be considered minimal. However, due to the differences in MUFA, PUFA and PUFA:SFA, it would have been interesting to observe the shelf stability of these two products, but this could not be done due to limitations of this study, thus further research on this aspect is recommended.

3.4. Descriptive Sensory Analysis

The sensory profiles of the two cabanossi treatments are presented in Table 5. There were significant differences for all aroma attributes ($p \leq 0.01$) except peppery aroma ($p > 0.05$). The PF treatment was scored higher for most of the aroma attributes. During the sensory panel training, some unique characteristics were detected in the SF cabanossi. Consequently, it was expected that this treatment would score higher for these characteristics viz., charred aroma, sheep-like fatty aroma, mutton aroma and herbaceous aroma ($p \leq 0.01$). The presence of these sensory attributes could be as a result of mutton specific heterocyclic compounds such as 2-ethyl-3,6-dimethylpyrazine and 2-pentylpyridine, as well as branched chain volatile fatty acids (BCFA) such as 4-methlyphenol acid, 4-methylnonanoic acid and 3-methyl-indole acid, commonly known as skatole [69]. There is a strong link between these BCFA and mutton aromas and flavour [70–72]. In fact, 4-methlyphenol, 4-methylnonanoic and 3-methyl-indole are BCFAs thought to be precursors of undesirable rancid odour and flavour in mutton [70]. Although the origin of these compounds is not clearly understood, they could be products of rumen metabolism of pasture species in the sheep diet [69,70]. Skatole, however, is known to be one of the major compounds that cause boar taint in *Suidae* spp. [69], thus the absence of these sensory attributes in PF cabanossi could be a good indication of the absence of boar taint in both pork backfat and warthog meat. Therefore, the hypothesis that these sensory attributes are SF-related is strengthened.

Pork backfat cabanossi tasted saltier ($p \leq 0.05$) than SF cabanossi although the salt content (~2.7%) did not differ ($p > 0.05$) in the chemical analysis (Table 3). This is attributed to the PF cabanossi being less fatty (chemically) with more protein. The bond between chloride ions and meat proteins is stronger than that of chloride and sodium ions (Hamm [73] cited in [74]), therefore, the extent to which chloride ions bind to protein may be strong enough to suppress the perception of salty flavour [74]. Regarding fatty mouthfeel, it was expected that SF cabanossi would score higher than PF cabanossi since it was higher in chemically analysed fat. Fat particles produce an oily coating around meat particles and this phenomenon causes a higher impression of fat upon chewing the cabanossi.

Concerning appearance, there were no differences ($p > 0.05$) in perceived percentage fat and fatty/oily/shininess, but the red-brown colour intensity differed between the two treatments. Red brown colour intensity was higher in PF cabanossi compared to SF cabanossi. This may be the influence of more protein recorded for PF cabanossi as opposed to more fat in the SF cabanossi. Fat is usually lighter in colour, thus, if more of it is present in a product, it will mask some of the dark/red colour of meat proteins.

Texture attributes were all significantly different between the two cabanossi products. Pork backfat cabanossi scored higher for first bite, chewiness and residue, whereas this was not the case for sustained juiciness. This is attributed to differences in protein and fat content of the products. Fat reduces hardness of meat and meat products by facilitating the diffusion of moisture during biting and mastication of meat [75]. Less energy and force are therefore required to successfully chew the product resulting in less residue. Furthermore, high fat meat sausages exhibit a better impression of juiciness compared to low fat sausages [75].

Figure 1 is a principal component analysis (PCA) showing the variation and grouping between the two cabanossi products. Factor 1 accounted for the most variation between the cabanossi. The PCA shows a clear distinction between the two products, with each being associated with specific attributes. The SF cabanossi was more associated with sheep-like fatty aroma and flavour, mutton fat aroma and flavour, herbaceous aroma and flavour, charred aroma as well as sustained juiciness. On the other hand, PF cabanossi was more associated with various aroma and flavour attributes typically linked to pork products as shown on the PCA.

Table 5. Sensory attributes of cabanossi made with either pork backfat (PF) or sheep tail/backfat (SF).

AROMA	PF	SF	*p*-Value
Overall aroma intensity	66.28 ± 1.85	56.66 ± 2.28	0.002
Smoky aroma	63.47 ± 1.77	52.34 ± 2.64	0.001
Charred aroma	4.94 ± 1.65	16.75 ± 1.92	0.001
Fermented sour aroma	19.27 ± 1.59	13.86 ± 1.41	0.006
Cured pork aroma	25.94 ± 1.97	15.44 ± 1.62	0.000
Pork fat aroma	17.10 ± 2.00	4.40 ± 1.79	0.000
Mutton aroma	1.02 ± 0.96	8.68 ± 1.29	0.000
Sheep-like fatty aroma	1.05 ± 0.98	8.67 ± 0.97	0.000
Peppery aroma	15.51 ± 0.70	14.86 ± 0.93	0.554
Sweet aroma	18.77 ± 0.42	14.14 ± 0.81	0.000
Herbaceous aroma	0.86 ± 0.12	3.99 ± 1.02	0.000
FLAVOUR			
Overall flavour intensity	63.03 ± 1.60	54.79 ± 1.45	0.000
Smoky flavour	57.27 ± 1.52	45.92 ± 2.43	0.000
Woody flavour	18.22 ± 0.96	12.75 ± 0.97	0.000
Cured pork flavour	34.54 ± 1.88	22.92 ± 1.50	0.000
Fermented sour flavour	23.05 ± 0.71	16.17 ± 1.74	0.000
Pork fat flavour	18.27 ± 2.05	3.98 ± 1.87	0.000
Mutton flavour	1.32 ± 1.47	21.41 ± 2.18	0.000
Sheep-like fatty flavour	1.52 ± 0.64	15.47 ± 1.70	0.000
Peppery flavour	17.15 ± 0.99	15.86 ± 0.87	0.176
Herbaceous flavour	2.17 ± 0.59	8.32 ± 1.22	0.000
Salty taste	19.72 ± 0.75	18.58 ± 0.74	0.031
Sweet taste	19.61 ± 1.27	16.13 ± 1.37	0.002
Fatty mouthfeel	16.17 ± 1.07	22.03 ± 1.14	0.000
APPEARANCE			
Red/brown colour intensity	52.89 ± 2.73	47.85 ± 4.46	0.027
Fatty/Oily/Shininess	48.95 ± 3.99	47.84 ± 4.65	0.991
Perceived Percentage fat	46.09 ± 1.00	44.57 ± 1.72	0.642
TEXTURE			
First bite	29.91 ± 1.63	28.67 ± 2.00	0.218
Sustained juiciness	45.16 ± 1.24	50.74 ± 2.46	0.014
Chewiness	25.42 ± 0.70	22.29 ± 0.89	0.002
Residue	23.12 ± 1.54	18.21 ± 1.08	0.000

All data are expressed as mean ± SE (n = 12).

3.5. Consumer Preference

In order to accurately predict consumer behaviour and attitudes towards new food products, it is important to understand various aspects of the population including their preference, choice, desire to eat certain foods, purchase intent and frequency of consumption [76]. Population demographic characteristics are a known influential factor on the sensory acceptance of healthier, reformulated meat products [76–78]. Results from the study revealed that most people consumed meat frequently (Figure 2). The majority of the population (62.5%) consume meat on a daily basis, whilst 25% consumed meat more than three times per week and only 12.5% of the population consumed meat between 1–3 times per week. However, game meat was not frequently consumed, with the majority (44.7%) of the population indicating that they only ate it approximately four times a year, whereas 36.8% of the population attest to consuming game meat at least twice a month. Whilst level of ethnicity and education were not included in the analyses due to statistical imbalances, gender and age group influenced ($p > 0.05$) the frequency of consumption of neither domestic nor game meat. Burger [79] reported that the consumption of game meat in North America was influenced by demographic characteristics such as ethnicity, gender and household income. Other studies such as that of Hoffman [32] also reported an association between ethnicity and game meat consumption in South Africa as some population groups associated it with leanness and healthiness, as well as it having a favourable gamey flavour which was not perceived as important by others. Game meat consumption may also be influenced by the lifestyles and social activities of different populations as families that are involved in hunting are more likely to regularly consume game meat [80]. However, a recent study did not report any associations between demography and game meat consumption [33].

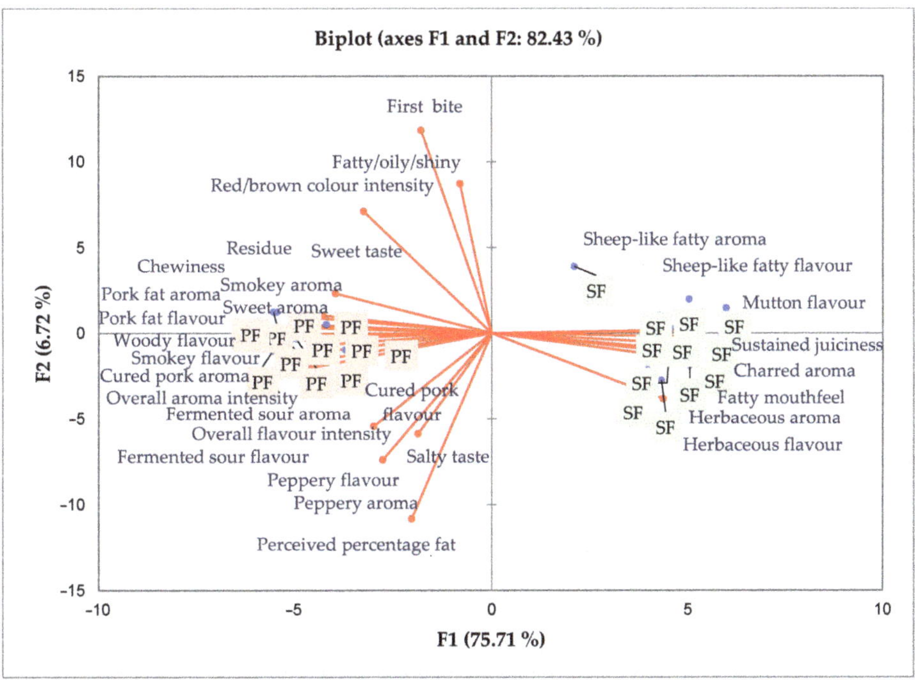

Figure 1. Principal component analysis for the sensory attributes of cabanossi made with either pork backfat (PF) or fat-tailed sheep fat (SF).

Consumers were asked to rank their preference (on a scale of 1 to 9; the higher the number the more positive their preference) for various game meat products in addition to factors that influence their purchasing decisions for game meat products. Least significant means for these rankings are shown in Table 6. In the order of preference, biltong (rating of 7.7) was the most preferred product followed by droëwors (6.9), fresh meat (6.7) and fresh/raw sausage (6.6), salami (6.3) and cabanossi (6.3). These results indicate that game meat might be more preferred if marketed as processed meat products rather than fresh meat as supported by the literature [81]. Consumers perceive fresh game meat to be difficult to prepare [32,81], probably due to limited knowledge on preparation methods, and would therefore prefer to consume it processed.

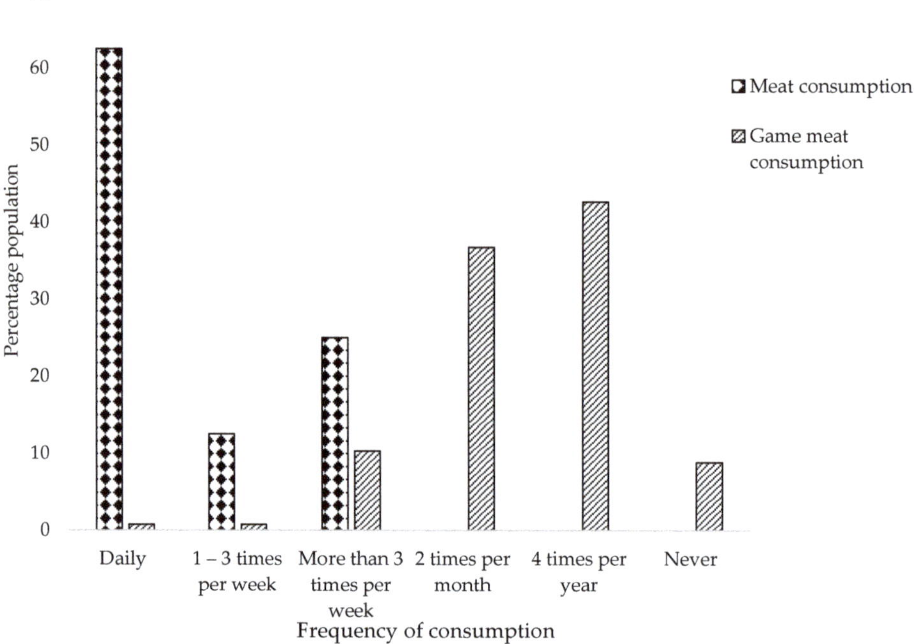

Figure 2. Meat consumption behaviour within the sample population.

Table 6. Least significant means (±SE) of the preference and factors influencing the purchase of game meat products within the sample population.

Product	Least Significant Mean	Factor	Least Significant Mean
Biltong	7.7 [a] ± 1.69	Availability	6.2 [c,d] ± 1.91
Cabanossi	6.3 [d] ± 2.01	Fat content	6.5 [b,c] ± 1.87
Droëwors	6.9 [b] ± 1.84	Origin	5.5 [e] ± 2.52
Fresh meat	6.7 [b,c] ± 1.89	Price	7.2 [a] ± 1.77
Fresh/raw sausage	6.6 [b] ± 1.81	Safety	6.7 [b] ± 2.16
Salami	6.3 [c,d] ± 1.94	Species	5.8 [d,e] ± 2.26

[a–e] Means with different superscripts between columns are significantly different.

Results from this study suggest that biltong was the most preferred game meat product among the sample population. This could be attributed to the fact that biltong has a strong linkage to the South African tradition as a meat preservation strategy that has long been known [35]. Furthermore, in South Africa, biltong is produced by small artisanal (e.g., households and butcheries) to large commercial manufactures [35], and thus most consumers who participated in the study could have developed a preference for this product at some point during their upbringing. Some of these consumers could have had experience in making biltong at home, whilst others got exposure to it in local butcheries and retail outlets. Therefore, background knowledge of a product could have effects on its acceptance as a desirable food. Henceforth, in the current investigation, it was interesting to note that contrary to the suggestions that consumers preferred processed game meat compared to fresh meat, consumers actually preferred fresh game meat compared to cabanossi and salami (Table 6). Since cabanossi and salami have Polish and Italian origins, respectively, they may not have been popular products among the sample population, which could have contributed to less preference. Consumer acceptance of different meat products is often a complex phenomenon encompassing several factors which include psychological and demographic factors as well as food choice habits [77]. Less preference reported for cabanossi and salami in this study could also be attributed to their high price since price was the single most important factor (7.2) affecting purchasing decisions (Table 6), especially if the product did not offer any known benefits to the consumer. Processed meat products are expected to fetch higher prices per kilogram unit because of a greater resource input. Therefore, these products must be produced to provide benefits beyond basic nutrition as functional meat products [82] if their preference among South African groups is to increase. The second most important factors determining the purchase of game meat products were fat content (6.5) and safety (6.7), while species (5.8) and origin (5.5) were ranked as the least contributing factors. Consumer trends are shifting towards eating healthy foods with reduced fat content to limit the onset of cardiovascular diseases, certain cancers and other food-related complications including microbial poisoning [82,83], although purchasing decisions for these meat products are governed by the availability of disposable income [77].

Least significant means for overall taste and appearance scores for the two-cabanossi treatments under current investigation are shown in Table 7. There were significant consumer preference differences ($p \leq 0.01$) pertaining to taste and appearance between the two cabanossi treatments. The PF cabanossi was ranked higher for appearance followed by the SF treatment. The appearance of the product produces the first impression that consumers will judge it by. Consumers use this impression to estimate product freshness, quality and probable sensory characteristics [84]. The observation that PF cabanossi was more preferred in terms of appearance could be linked with higher ratings for red/brown colour intensity observed in the DSA. Consumers prefer darker, redder sausages with less perceived percentage fat because they consider them healthier meat products [85]. Although the perceived percentage fat was similar in both treatments during DSA, during consumer analysis, it was observed that SF cabanossi had an external oily sheen which could have been caused by pressure exerted on the sausages during vacuum packaging. Similarly, the observation that PF cabanossi scored higher for taste is in accordance with the DSA, which found that it received higher scores for the most favourable flavour attributes including overall flavour intensity, smoky flavour, cured pork flavour, peppery flavour and sweet taste (Table 5). Generally, both treatments were rendered acceptable by the consumers, receiving scores of more than 6. The consumers' acceptance of cabanossi from this study suggests that it is possible to produce acceptable cabanossi using SF.

Table 7. Least significant means for overall acceptance of cabanossi made with either pork backfat (PF) or sheep tail/backfat (SF).

Attribute	PF	SF
Appearance	6.75 [a] ± 1.38	6.27 [b] ± 1.61
Taste	6.75 [a] ± 1.64	6.12 [b] ± 1.81

[a,b] Means with different superscripts between columns are significantly different. All data are expressed as mean ± SE.

4. Conclusions

Utility of different types of fat affects some physicochemical and sensorial characteristics of meat products. The findings obtained from this study indicate that fat-tailed sheep tail and backfat can be used as an alternative to pork backfat without detrimental effects on the physicochemical characteristics of cabanossi. However, results from descriptive sensory analysis indicate that this replacement produces products that are far apart from each other concerning aroma, flavour and texture, although, according to the ratings within the scales used, both products are acceptable. Although PF cabanossi were scored higher for most sensory attributes, SF cabanossi had some unique pleasant sensory attributes that are acceptable to consumers. Therefore, fat-tailed sheep tail and backfat may be used not necessarily as a replacement for pork backfat, but to produce another variety of cabanossi to diversify consumer choices. It may be necessary, and therefore, recommended that a shelf-life study be conducted to determine the influence of SF on shelf stability and flavour compounds of cabanossi. Valorisation of fat-tailed sheep breeds fat to develop new meat products that may be useful in improving income for artisanal meat product manufactures through product diversification. However, this may be dependent on region/country as people from areas where eating pork is not acceptable may be more receptive of sheep meat aroma and flavour in their meat products.

Author Contributions: Conceptualization and methodology, L.N.M., M.R., E.A. and L.C.H.; funding acquisition, V.M. and L.C.H.; writing—original draft preparation, L.N.M.; writing—review and editing, E.A. and L.C.H.; supervision, M.R., E.A., V.M. and L.C.H. All authors have read and agreed to the published version of the manuscript.

Funding: This research is supported by the South African Research Chairs Initiative (SARChI) and partly funded by the South African Department of Science and Technology (UID number: 84633), as administered by the National Research Foundation (NRF) of South Africa, and partly by the Department of Trade and Industry's THRIP program (THRIP/64/19/04/2017) with Wildlife Ranching South Africa as a partner. Any opinion, finding and conclusion or recommendation expressed in this material is that of the author(s) and the NRF does not accept any liability whatsoever in this regard.

Acknowledgments: Sincere gratitude goes to the technical and administrative staff from the Departments of Animal Sciences and Food Science, Stellenbosch University, South Africa and Department of Livestock and Pasture Science, University of Fort Hare, South Africa. All the contributions of the late Voster Muchenje will forever be cherished. May his dear soul continue to rest in eternal peace.

Conflicts of Interest: The authors declare no conflict of interest. The funders had no role in the design of the study; in the collection, analyses, or interpretation of data; in the writing of the manuscript, or in the decision to publish the results.

References

1. Jiménez-Colmenero, F.; Reig, M.; Toldrá, F. New approaches for the development of functional meat products. In *Advanced Technologies For Meat Processing*; CRC Press: Boca Raton, FL, USA, 2006; pp. 275–308. ISBN 978-1-4200-1731-1.

2. Chikwanha, O.C.; Vahmani, P.; Muchenje, V.; Dugan, M.E.R.; Mapiye, C. Nutritional enhancement of sheep meat fatty acid profile for human health and wellbeing. *Food Res. Int.* **2018**, *104*, 25–38. [CrossRef] [PubMed]

3. Ponnampalam, E.N.; Sinclair, A.J.; Egan, A.R.; Ferrier, G.R.; Leury, B.J. Dietary manipulation of muscle long-chain omega-3 and omega-6 fatty acids and sensory properties of lamb meat. *Meat Sci.* **2002**, *60*, 125–132. [CrossRef]

4. Vasta, V.; Nudda, A.; Cannas, A.; Lanza, M.; Priolo, A. Alternative feed resources and their effects on the quality of meat and milk from small ruminants. *Anim. Feed Sci. Technol.* **2008**, *147*, 223–246. [CrossRef]

5. Baer, A.A.; Dilger, A.C. Effect of fat quality on sausage processing, texture, and sensory characteristics. *Meat Sci.* **2014**, *96*, 1242–1249. [CrossRef]

6. Mordenti, A.L.; Martelli, G.; Brogna, N.; Nannoni, E.; Vignola, G.; Zaghini, G.; Sardi, L. Effects of a soybean-free diet supplied to Italian heavy pigs on fattening performance, and meat and dry-cured ham quality. *Ital. J. Anim. Sci.* **2012**, *11*, 459–465. [CrossRef]

7. Della Casa, G.; Bochicchio, D.; Faeti, V.; Marchetto, G.; Poletti, E.; Rossi, A.; Panciroli, A.; Mordenti, A.L.; Brogna, N. Performance and fat quality of heavy pigs fed maize differing in linoleic acid content. *Meat Sci.* **2010**, *84*, 152–158. [CrossRef]

8. Yousefi, A.R.; Kohram, H.; Zare Shahneh, A.; Nik-khah, A.; Campbell, A.W. Comparison of the meat quality and fatty acid composition of traditional fat-tailed (Chall) and tailed (Zel) Iranian sheep breeds. *Meat Sci.* **2012**, *92*, 417–422. [CrossRef]

9. Wood, J.D.; Richardson, R.I.; Nute, G.R.; Fisher, A.V.; Campo, M.M.; Kasapidou, E.; Sheard, P.R.; Enser, M. Effects of fatty acids on meat quality: A review. *Meat Sci.* **2003**, *66*, 21–32. [CrossRef]

10. Ospina-E, J.C.; Sierra-C, A.; Ochoa, O.; Pérez-Álvarez, J.A.; Fernández-López, J. Substitution of saturated fat in processed meat products: A review. *Crit. Rev. Food Sci. Nutr.* **2012**, *52*, 113–122. [CrossRef]

11. Sasaki, K.; Mitsumoto, M.; Nishioka, T.; Irie, M. Differential scanning calorimetry of porcine adipose tissues. *Meat Sci.* **2006**, *72*, 789–792. [CrossRef] [PubMed]

12. Świątkiewicz, M.; Oczkowicz, M.; Ropka-Molik, K.; Hanczakowska, E. The effect of dietary fatty acid composition on adipose tissue quality and expression of genes related to lipid metabolism in porcine livers. *Anim. Feed Sci. Technol.* **2016**, *216*, 204–215. [CrossRef]

13. Cunha, L.C.M.; Lúcia, M.; Monteiro, G.; Lorenzo, J.M.; Munekata, P.E.S.; Muchenje, V.; Allan, F.; De Carvalho, L.; Conte-junior, C.A. Natural antioxidants in processing and storage stability of sheep and goat meat products. *Food Res. Int.* **2018**, *111*, 379–390. [CrossRef] [PubMed]

14. Hugo, A.; Roodt, E. Significance of porcine fat quality in meat technology: A review. *Food Rev. Int.* **2007**, *23*, 175–198. [CrossRef]

15. Ščetar, M.; Kurek, M.; Galić, K. Trends in meat and meat products packaging—A review. *Croat. J. Food Sci. Technol.* **2010**, *2*, 32–48.

16. Stasiewicz, M.; Lipiński, K.; Cierach, M. Quality of meat products packaged and stored under vacuum and modified atmosphere conditions. *J. Food Sci. Technol.* **2014**, *51*, 1982–1989. [CrossRef]

17. Estevez, M. Oxidative damage to poultry: From farm to fork. *Poult. Sci.* **2015**, *94*, 1368–1378. [CrossRef]

18. Mariutti, L.R.B.; Bragagnolo, N. Influence of salt on lipid oxidation in meat and seafood products: A review. *Food Res. Int.* **2017**, *94*, 90–100. [CrossRef]

19. Arnaud, E.; Santchurn, S.; Collignan, A. Fermented Poultry Sausages. In *Handbook of Fermented Meat and Poultry*; Toldrá, F., Hui, Y.H., Astiasaran, I., Nip, W.-K., Sebranek, J.G., Silveira, E.-T., Talon, R., Eds.; Blackwell Publishing Ltd.: Iowa City, IA, USA, 2015.

20. Ruiz, J. Ingredients. In *Handbook of Fermented Meat and Poultry*; Toldrá, F., Hui, Y.H., Astiasaran, I., Nip, W.-K., Sebranek, J.G., Silveira, E.-T., Talon, R., Eds.; Blackwell Publishing Ltd.: Iowa City, IA, USA, 2007; pp. 59–76.

21. Kashan, N.E.J.; Manafi Azar, G.H.; Afzalzadeh, A.; Salehi, A. Growth performance and carcass quality of fattening lambs from fat-tailed and tailed sheep breeds. *Small Rumin. Res.* **2005**, *60*, 267–271. [CrossRef]

22. Moharrery, A. Effect of docking and energy of diet on carcass fat characteristics in fat-tailed Badghisian sheep. *Small Rumin. Res.* **2007**, *69*, 208–216. [CrossRef]

23. Khaldari, M.; Kashan, N.E.J.; Afzalzadeh, A.; Salehi, A. Growth and carcass characteristics of crossbred progeny from lean tailed and fat tailed sheep breeds. *S. Afr. J. Anim. Sci.* **2007**, *37*, 51–56. [CrossRef]

24. Wang, Y.Q.; Zhong, R.Z.; Fang, Y.; Zhou, D.W. Influence of tail docking on carcass characteristics, meat quality and fatty acid composition of fat-tail lambs. *Small Rumin. Res.* **2018**, *162*, 17–21. [CrossRef]

25. Alves, S.P.; Bessa, R.J.B.; Quaresma, M.A.G.; Kilminster, T.; Scanlon, T.; Oldham, C.; Milton, J.; Greeff, J.; Almeida, A.M. Does the Fat Tailed Damara Ovine Breed Have a Distinct Lipid Metabolism Leading to a High Concentration of Branched Chain Fatty Acids in Tissues? *PLoS ONE* **2013**, *8*, e77313. [CrossRef] [PubMed]

26. van Harten, S.; Kilminster, T.; Scanlon, T.; Milton, J.; Oldham, C.; Greeff, J.; Almeida, A.M. Fatty acid composition of the ovine longissimus dorsi muscle: Effect of feed restriction in three breeds of different origin. *J. Sci. Food Agric.* **2016**, *96*, 1777–1782. [CrossRef] [PubMed]

27. North, M.K.; Dalle Zotte, A.; Hoffman, L.C. The use of dietary flavonoids in meat production: A review. *Anim. Feed Sci. Technol.* **2019**, *257*, 114291. [CrossRef]

28. Gagaoua, M.; Boudechicha, H.R. Ethnic meat products of the North African and Mediterranean countries: An overview. *J. Ethn. Foods* **2018**, *5*, 83–98. [CrossRef]

29. Teixeira, A.; Silva, S.; Guedes, C.; Rodrigues, S. Sheep and Goat Meat Processed Products Quality: A Review. *Foods* **2020**, *9*, 960. [CrossRef]

30. Çiçek, Ü.; Polat, N. Investigation of physicochemical and sensorial quality of a type of traditional meat product: Bez sucuk. *LWT—Food Sci. Technol.* **2016**, *65*, 145–151. [CrossRef]

31. Jones, M.; Hoffman, L.C.; Muller, M. Effect of rooibos extract (*Aspalathus linearis*) on lipid oxidation over time and the sensory analysis of blesbok (*Damaliscus pygargus phillipsi*) and springbok (*Antidorcas marsupialis*) droëwors. *Meat Sci.* **2015**, *103*, 54–60. [CrossRef]

32. Hoffman, L.C.; Muller, M.; Schutte, D.W.; Calitz, F.J.; Crafford, K. Consumer expectations, perceptions and purchasing of South African game meat. *S. Afr. J. Wildl. Res.* **2005**, *35*, 33–42.

33. Swanepoel, M.; Leslie, A.J.; Hoffman, L.C. Comparative analyses of the chemical and sensory parameters and consumer preference of a semi-dried smoked meat product (cabanossi) produced with warthog (*Phacochoerus africanus*) and domestic pork meat. *Meat Sci.* **2016**, *114*, 103–113. [CrossRef]

34. Chakanya, C.; Arnaud, E.; Muchenje, V.; Hoffman, L.C. Fermented meat sausages from game and venison: What are the opportunities and limitations? *J. Sci. Food Agric.* **2017**. [CrossRef]

35. Jones, M.; Arnaud, E.; Gouws, P.; Hoffman, L.C. Processing of South African biltong—A review. *S. Afr. J. Anim. Sci.* **2017**, *47*, 743–757. [CrossRef]

36. Mahachi, L.N.; Rudman, M.; Arnaud, E.; Muchenje, V.; Hoffman, L.C. Development of semi dry sausages (cabanossi) with warthog (*Phacochoerus africanus*) meat: Physicochemical and sensory attributes. *LWT—Food Sci. Technol.* **2019**, *115*, 108454. [CrossRef]

37. Tyburcy, A.; Kozyra, D. Effects of composite surface coating and pre-drying on the properties of kabanosy dry sausage. *Meat Sci.* **2010**, *86*, 405–410. [CrossRef] [PubMed]

38. Law in the EU. *Eur. Food Feed Law Rev.* **2011**, *6*, 351–353.

39. Chmiel, M.; Adamczak, L.; Wronska, K.; Pietrzak, D.; Florowski, T. The effect of drying parameters on the quality of pork and poultry-pork kabanosy produced according to the traditional specialties guaranteed recipe. *J. Food Qual.* **2017**, 1597432. [CrossRef]

40. Swanepoel, M.; Leslie, A.J.; van der Rijst, M.; Hoffman, L.C. Physical and chemical characteristics of warthog (*Phacochoerus africanus*) meat. *Afr. J. Wildl. Res.* **2016**, *46*, 103–120. [CrossRef]

41. Hoffman, L.C.; Sales, J. Physical and chemical quality characteristics of warthog (*Phacochoerus aethiopicus*) meat. *Livest. Res. Rural Dev.* **2007**, *19*, 153.

42. Rudman, M.; Leslie, A.J.; van der Rijst, M.; Hoffman, L.C. Quality characteristics of warthog (*Phacochoerus africanus*) meat. *Meat Sci.* **2018**, *145*, 266–272. [CrossRef]

43. AOAC. *Official Methods of Analysis*, 18th ed.; Association of Official Analytical Chemists: Washington, DC, USA, 2006.

44. Lee, C.M.; Trevino, B.; Chaiyawat, M. A simple and rapid solvent extraction method for determining total lipids in fish tissue. *J. AOAC Int.* **1996**, *79*, 487–492. [CrossRef]

45. Neethling, J.; Britz, T.J.; Hoffman, L.C. Impact of season on the fatty acid profiles of male and female blesbok (*Damaliscus pygargus phillipsi*) muscles. *Meat Sci.* **2014**, *98*, 599–606. [CrossRef] [PubMed]

46. Ulbricht, T.L.V.; Southgate, D.A.T. Coronary heart disease: Seven dietary factors. *Lancet* **1991**, *338*, 985–992. [CrossRef]

47. Mukumbo, F.E.; Arnaud, E.; Collignan, A.; Hoffman, L.C.; Descalzo, A.M.; Muchenje, V. Physico-chemical composition and oxidative stability of South African beef, game, ostrich and pork droëwors. *J. Food Sci. Technol.* **2018**, *55*, 4833–4840. [CrossRef] [PubMed]

48. Lawless, H.T.; Heimann, H. *Sensory Evaluation of Food*, 2nd ed.; Springer: New York, NY, USA, 2010.

49. Ekine-Dzivenu, C.; Chen, L.; Vinsky, M.; Aldai, N.; Dugan, M.E.R.; McAllister, T.A.; Wang, Z.; Okine, E.; Li, C. Estimates of genetic parameters for fatty acids in brisket adipose tissue of Canadian commercial crossbred beef steers. *Meat Sci.* **2014**, *96*, 1517–1526. [CrossRef]

50. Mora-Gallego, H.; Serra, X.; Guàrdia, M.D.; Miklos, R.; Lametsch, R.; Arnau, J. Effect of the type of fat on the physicochemical, instrumental and sensory characteristics of reduced fat non-acid fermented sausages. *Meat Sci.* **2013**, *93*, 668–674. [CrossRef]

51. Tyburcy, A.; Wasiak, P.; Cegiełka, A. Application of composite protective coatings on the surface of sausages with different water content. *Acta Sci. Pol. Technol. Aliment.* **2010**, *9*, 151–159.

52. McQuestin, O.J.; Shadbolt, C.T.; Ross, T. Quantification of the relative effects of temperature, pH, and water activity on inactivation of Escherichia coli in fermented meat by meta-analysis. *Appl. Environ. Microbiol.* **2009**, *75*, 6963–6972. [CrossRef]

53. Toldrá, F. Biochemistry of fermented meat. In *Food Biochemistry and Food Processing*; Simpson, B.K., Nollet, L.M., Toldrá, F., Benjakul, S., Paliyath, G., Hui, Y.H., Eds.; Johnn Wiley & Sons, Inc.: Oxford, UK, 2012; pp. 331–343.

54. Ross, T.; Shadbolt, C.T. *Predicting Escherichia coli Inactivation in Uncooked Comminuted Fermented Meat Products*; Meat and Livestock Australia: Sydney, Australia, 2001; Volume 364.

55. Muguerza, E.; Fista, G.; Ansorena, D.; Astiasaran, I.; Bloukas, J.G. Effect of fat level and partial replacement of pork backfat with olive oil on processing and quality characteristics of fermented sausages. *Meat Sci.* **2002**, *61*, 397–404. [CrossRef]

56. Gómez, M.; Lorenzo, J.M. Effect of fat level on physicochemical, volatile compounds and sensory characteristics of dry-ripened "chorizo" from Celta pig breed. *Meat Sci.* **2013**, *95*, 658–666. [CrossRef]

57. Ammor, M.S.; Mayo, B. Selection criteria for lactic acid bacteria to be used as functional starter cultures in dry sausage production: An update. *Meat Sci.* **2007**, *76*, 138–146. [CrossRef]

58. Lorenzo, J.M.; Fontan, M.C.G.; Cachaldora, A.; Franco, I.; Carballo, J. Study of the lactic acid bacteria throughout the manufacture of dry-cured lacón (a Spanish traditional meat product). Effect of some additives. *Food Microbiol.* **2010**, *27*, 229–235. [CrossRef] [PubMed]
59. Hoffman, L.C.; Jones, M.; Muller, N.; Joubert, E.; Sadie, A. Lipid and protein stability and sensory evaluation of ostrich (*Struthio camelus*) droewors with the addition of rooibos tea extract (*Aspalathus linearis*) as a natural antioxidant. *Meat Sci.* **2014**, *96*, 1289–1296. [CrossRef] [PubMed]
60. Yim, D.G.; Jang, K.H.; Chung, K.Y. Effect of fat level and the ripening time on quality traits of fermented sausages. *Asian-Australas. J. Anim. Sci.* **2016**, *29*, 119–125. [CrossRef] [PubMed]
61. Warris, P. *Meat Science*; CAB Publishing: Oxon, UK, 2000.
62. Lorenzo, J.M.; Montes, R.; Purriños, L.; Cobas, N.; Franco, D. Fatty acid composition of Celta pig breed as influenced by sex and location of fat in the carcass. *J. Sci. Food Agric.* **2012**, *92*, 1311–1317. [CrossRef] [PubMed]
63. Valencak, T.G.; Gamsjäger, L.; Ohrnberger, S.; Culbert, N.J.; Ruf, T. Healthy n-6/n-3 fatty acid composition from five European game meat species remains after cooking. *BMC Res. Notes* **2015**, *8*, 273. [CrossRef]
64. Wood, J.D.; Enser, M.; Fisher, A.V.; Nute, G.R.; Sheard, P.R.; Richardson, R.I.; Hughes, S.I.; Whittington, F.M. Fat deposition, fatty acid composition and meat quality: A review. *Meat Sci.* **2008**, *78*, 343–358. [CrossRef]
65. Romero, M.C.; Romero, A.M.; Doval, M.M.; Judis, A. Nutritional value and fatty acid composition of some traditional Argentinean meat sausages. *Food Sci. Technol.* **2013**, *33*, 161–166. [CrossRef]
66. Simopoulos, A.P. An increase in the Omega-6/Omega-3 fatty acid ratio increases the risk for obesity. *Nutrients* **2016**, *8*, 128. [CrossRef]
67. Del Nobile, M.A.; Conte, A.; Incoronato, A.L.; Panza, O.; Sevi, A.; Marino, R. New strategies for reducing the pork back-fat content in typical Italian salami. *Meat Sci.* **2009**, *81*, 263–269. [CrossRef]
68. Mukumbo, F.E.; Descalzo, A.M.; Collignan, A.; Hoffman, L.C.; Servent, A.; Muchenje, V.; Arnaud, E. Effect of Moringa oleifera leaf powder on drying kinetics, physico-chemical properties, ferric reducing antioxidant power, α-tocopherol, β-carotene, and lipid oxidation of dry pork sausages during processing and storage. *J. Food Process. Preserv.* **2020**, *44*, 1–12. [CrossRef]
69. Channon, H.A.; Lyons, R.; Bruce, H. Sheepmeat Flavour and Odour: A Review. Sheep CRC Project Number Sheep CRC 1.3.2. 2003. Available online: http://www.sheepcrc.org.au/images/pdfs/CRC1/CRC1_Meat/SMEQ/Sheepmeat_flavour_review.pdf (accessed on 8 November 2016).
70. Young, O.A.; Berdague, J.-L.; Viallon, C.; Rousset-Akrim, S.; Theriez, M. Fat-borne volatiles and sheepmeat odour. *Meat Sci.* **1997**, *45*, 183–200. [CrossRef]
71. Watkins, P.J.; Frank, D. Heptadecanoic acid as an indicator of BCFA content in sheep fat. *Meat Sci.* **2019**, *151*, 33–35. [CrossRef] [PubMed]
72. Watkins, P.J.; Kearney, G.; Rose, G.; Allen, D.; Ball, A.J.; Pethick, D.W.; Warner, R.D. Effect of branched-chain fatty acids, 3-methylindole and 4-methylphenol on consumer sensory scores of grilled lamb meat. *Meat Sci.* **2014**, *96*, 1088–1094. [CrossRef]
73. Hamm, R. *Kolloidchemie des Fleisches*; Parey: Berlin/Hamburg, Germany, 1972.
74. Ruusunen, M.; Simolin, M.; Puolanne, E. The effect of fat content and flavor enhancers on the perceived saltiness of cooked "bologna-type" sausages. *J. Muscle Foods* **2001**, *27*, 107–120. [CrossRef]
75. Lorenzo, J.M.; Temperán, S.; Bermúdez, R.; Purriños, L.; Franco, D. Effect of fat level on physicochemical and sensory properties of dry-cured duck sausages. *Poult. Sci.* **2011**, *90*, 1334–1339. [CrossRef] [PubMed]
76. Guiné, R.P.F.; Florença, S.G.; Barroca, M.J.; Anjos, O. The Link between the Consumer and the Innovations in Food Product Development. *Foods* **2020**, *9*, 1317. [CrossRef] [PubMed]
77. Shan, L.C.; De Brún, A.; Henchion, M.; Li, C.; Murrin, C. Consumer evaluations of processed meat products reformulated to be healthier—A conjoint analysis study. *Meat Sci.* **2017**, *131*, 82–89. [CrossRef]
78. Cardoso, A.P.; Ferreira, V.; Leal, M.; Ferreira, M.; Campos, S.; Guiné, R.P.F. Perceptions about healthy eating and emotional factors conditioning eating behaviour: A study involving Portugal, Brazil and Argentina. *Foods* **2020**, *9*, 1236. [CrossRef]
79. Burger, J. Daily consumption of wild fish and game: Exposures of high end recreationists. *Int. J. Environ. Health Res.* **2002**, *12*, 343–354. [CrossRef]
80. Radder, L.; le Roux, R. Factors affecting food choice in relation to venison: A South African example. *Meat Sci.* **2005**, 583–589. [CrossRef]

81. Hoffman, L.C.; Wiklund, E. Game and venison—Meat for the modern consumer. *Meat Sci.* **2006**, *74*, 197–208. [CrossRef]

82. Arihara, K. Strategies for designing novel functional meat products. *Meat Sci.* **2006**, *74*, 219–229. [CrossRef]

83. Jiménez-Colmenero, F.; Carballo, J.; Cofrades, S. Healthier meat and meat products: Their role as functional foods. *Meat Sci.* **2001**, *59*, 5–13. [CrossRef]

84. Font-i-Furnols, M.; Guerrero, L. Consumer preference, behavior and perception about meat and meat products: An overview. *Meat Sci.* **2014**, *98*, 361–371. [CrossRef]

85. Girolami, A.; Napolitano, F.; Faraone, D.; Di Bello, G.; Braghieri, A. Image analysis with the computer vision system and the consumer test in evaluating the appearance of Lucanian dry sausage. *Meat Sci.* **2014**, *96*, 610–616. [CrossRef]

Publisher's Note: MDPI stays neutral with regard to jurisdictional claims in published maps and institutional affiliations.

Article

Reduction of Salt and Fat in Frankfurter Sausages by Addition of *Agaricus bisporus* and *Pleurotus ostreatus* Flour

Magdalena I. Cerón-Guevara [1], Esmeralda Rangel-Vargas [1], José M. Lorenzo [2], Roberto Bermúdez [2], Mirian Pateiro [2], Jose A. Rodríguez [1], Irais Sánchez-Ortega [1] and Eva M. Santos [1,*

[1]   Universidad Autónoma del Estado de Hidalgo, Área Académica de Química, Crta. Pachuca-Tulancingo Km 4.5 s/n, Col. Carboneras, Mineral de la Reforma, HID 42183, Mexico; magdacerondcash@gmail.com (M.I.C.-G.); esmeralda_rangel10403@uaeh.edu.mx (E.R.-V.); josear@uaeh.edu.mx (J.A.R.); irais_sanchez5498@uaeh.edu.mx (I.S.-O.)

[2]   Meat Technology Centre of Galicia, Rúa Galicia N° 4, Parque Tecnológico de Galicia, San Cibrao das Viñas, 32900 Ourense, Spain; jmlorenzo@ceteca.net (J.M.L.); robertobermudez@ceteca.net (R.B.); mirianpateiro@ceteca.net (M.P.)

*   Correspondence: emsantos@uaeh.edu.mx; Tel.: +52-771-717-20-00 (ext. 2516)

Received: 12 May 2020; Accepted: 1 June 2020; Published: 9 June 2020

Abstract: The reduction of fat and salt and the incorporation of fiber-rich compounds in frankfurters is a trend to improve their nutritional profile. The objective of this study was to evaluate the partial replacement of 30 and 50% of pork backfat and 50% of salt by adding edible mushroom flour (2.5 and 5%) from *Agaricus bisporus* (Ab) and *Pleurotus ostreatus* (Po) on physicochemical, microbiological and sensory properties of frankfurters sausages during cold storage. The addition of flours increased the moisture, and the dietary fiber contents in frankfurters, keeping the amino acid profile. Lipid oxidation remained under acceptable values despite not antioxidant effect was observed by mushrooms flours. Only spore-forming bacteria were found during cold storage. Color and texture was modified by addition of mushroom, being the Ab samples darker, while Po flour addition resulted in softer and less cohesive sausages. Although lower color, flavor, and taste scores were given to the mushroom samples than the control, they ranked in the acceptable level confirming that the inclusion of 2.5 and 5% of Ab and Po flours in fat- and salt-reduced frankfurter sausages resulted a feasible strategy to enhance the nutritional profile these products.

Keywords: edible mushroom flour; *Agaricus bisporus*; *Pleurotus ostreatus*; healthier meat products

---

1. Introduction

Meat and meat products have excellent nutritional properties, being sources of protein, fat, essential amino acids, minerals and vitamins [1,2]. However, due to its higher lipid profile and sodium content, apart from other contaminants present after obtaining and processing, meat consumption has lately been connected to the development of different diseases around the world, mainly diabetes, obesity, cardiovascular problems and cancer [3]. Therefore, the increasing demand for healthier meat products has driven to the meat industry and scientific community to look for new meat products with less fat and salt content, with dietary fiber addition, and even probiotics inclusion, or natural antioxidants and vegetable proteins addition [4–6].

The reduction of fat or salt and substitution by other vegetable sources or ingredients to improve the functional and nutritional properties has been widely tested in emulsified sausages like frankfurters considering these sausages may contain up to 30% pork fat, 40% of which is saturated fatty acids [7,8].

However, since they are highly demanded and consumed, maintaining the sensory quality in the improvement of nutritional value is essential [9].

In this sense, different oil sources like olive, canola, linseed and fish oil as well as cereal and seed fibers (rice, chia, linseed) or vegetables, legumes and fruits, even algae, have been included in frankfurters considering the nutritional properties but also the technological issues associated to the emulsion formation [10–16]. One alternative to improve the nutritional profile of meat products could be the edible mushrooms since they have been reported to be a rich source of essential nutrients, with a high content in protein (16.47–36.96%), low level of lipids, and high dietary fiber content (24.4–46.62% in dry weight, DW) [17]. Because of their interesting flavor, medicinal and quality nutritional aspects, they have been used as food supplement to prevent malnutrition [18]. Apart from that, the strong umami taste of mushroom often referred to as a meaty taste [19,20] could contribute to reduce the salt content in meat formulations [21].

Moreover, cultivation of edible mushrooms is an easily and economically viable process, due to their ability to grow on different wastes such as cereal, cotton, fruit, vegetable, sawdust and leaf [22]. Several successful attempts to use mushroom flour in some bakery products have been reported [23–25] but less information is available about its incorporation in meat products with different results [8,26]. Due to the above reasons, mushrooms could be considered a promising naturally functional ingredient to improve nutritional quality of meat products, being *Agaricus bisporus* (Ab; champignon) and *Pleurotus ostreatus* (Po; oyster mushroom) the most worldwide cultivated species [27]. The aim of this investigation was evaluate the effect of partial replacement of fat and salt by addition of edible mushroom (Ab and Po) flour in frankfurter type sausages on the physicochemical, microbiological and sensorial parameters during cold storage.

2. Materials and Methods

2.1. Flour Preparation

Lots of 4 kg from Ab and Po mushroom species were purchased from a local market in Mineral de la Reforma (Hidalgo, Mexico). Mushrooms were carefully selected based on visual appearance (light color without visible damage) and then rinsed, drained and cut in 5 mm-thick slices. Pieces of Ab were immersed in an acetic acid solution (2%) at 80 °C for 10 min and water-cooled at 4 °C to prevent enzymatic reactions. After draining off the excess liquid, the samples were dried in air-recirculating oven (CE5F, Shel Lab, Cornelius, OR, USA) at 60 ± 3 °C for 18 h, milled in a UDY cyclone sample mill (UDY Corp., Fort Collins, CO, USA) and sleeved through a 0.5 mm mesh. The resultant flours were placed in hermetic polyethylene bags and stored in the dark at room temperature until use.

2.2. Manufacture of Frankfurter-Type Sausages

Frankfurters were elaborated in the Meat Technology Centre (Ourense, Spain). Five formulations were designed to reduce fat (30% and 50% reductions) and salt, phosphates and caseinate (50% reduction) by the addition of Ab and Po flours (2.5% and 5.0%), comparing with the control formulation (C) elaborated with 25% fat and 1.5% salt, 2% of sodium caseinate and 0.5% of phosphates (Table 1). MF-Ab and MF-Po were the codes to identify sausages with 30% of fat reduction (medium-fat) and 50% of salt-phosphates-caseinate reduction, and with 2.5% Ab and Po flours, respectively. LF-Ab and LF-Po were the codes used for the sausages with 50% fat reduction (low-fat) and 50% salt-phosphates-caseinate reduction, and 5.0% Ab and Po flour added, respectively. MF-PoAb were assigned to identify the sausages with 30% fat reduction and 50% salt-phosphates-caseinate reduction, incorporated with 5.0% Ab and Po mix flour (2.5% each one).

Before the manufacturing process, sodium chloride and sodium ascorbate were added to the meat cut in 1 cm cubes with a rest period of 2 h. The lean and fat were minced, using plates of 8 mm and 6 mm respectively in a refrigerated mincer machine (La Minerva, Bologna, Italy). Potato starch and 50% of total sodium caseinate (dissolved in water) were previously added to the meat. Then, the rest of ingredients with the fat were added and mixed to homogeneity in a chilled cutter (TALSA K30 0542, Talsabell S.A., Valencia, Spain). Meat batter was stuffed into 25 mm collagen casings with an automatic equipment (SIA, Junior, Barcelona, Spain) forming 10 cm frankfurters.

The raw sausages were cooked in a temperature-controlled bath-water (Marmite Mera REA-505, Talsa, Talsabell S.A., Valencia, Spain) at 90 °C for 20 min. Then, the cooked sausages were cooled in an ice water-bath. Four frankfurters were vacuum packed (FRIMAQ V900, Lorca, Spain) in each polyethylene bag, pasteurized at 90 °C for 30 min and kept at 2 °C storage until the laboratory analysis during the 90-days storage period. Four packages of frankfurters from each batch were taken at 0, 30, 60 and 90 days of storage. All analyses were carried out in triplicate for each formulation. Chemical composition and amino acid profile were evaluated only on day 0. After the sampling for microbial analysis, the sausages were oxygenated for 30 min at room temperature for the rest of analyses.

Table 1. Formulation (g) of different frankfurters with edible mushroom flour.

	Control [1]	MF-Ab	MF-Po	LF-Ab	LF-Po	MF-PoAb
Lean meat	2500	2500	2500	2500	2500	2500
Flour *Agaricus bisporus*	0	125	0	250	0	125
Flour *Pleurotus ostreatus*	0	0	125	0	250	125
Pork backfat	1250	875	875	625	625	875
Sodium caseinate	100	50	50	50	50	50
Sodium chloride	75	37.5	37.5	37.5	37.5	37.5
Water	1000	1346	1346	1471	1471	1471
Potato starch	50	50	50	50	50	50
Di-tri phosphates	25	12.5	12.5	12.5	12.5	12.5
Sodium nitrite	1.5	1.5	1.5	1.5	1.5	1.5
Sodium ascorbate	2.5	2.5	2.5	2.5	2.5	2.5

[1] Batches: Control (25% fat, 1.5% salt, 2% Sodium caseinate and 0.5% phosphates), MF-Ab (30% fat reduction, 50% salt reduction, 2.5% Ab), MF-Po (30% fat reduction, 50% salt reduction, 2.5% Po), LF-Ab (50% fat reduction, 50% salt reduction, 5.0% Ab), LF-Po (50% fat reduction, 50% salt reduction, 5.0% Po) and MF-PoAb (30% fat reduction, 50% salt reduction, 2.5% Ab and 2.5% Po).

2.3. Chemical Composition

Moisture, ash and protein were quantified according to the International Organisation for Standardisation (ISO) recommended standards 1442:1997 [28], 936:1998 [29] and 937:1978 [30], respectively. Crude fat was extracted using an Ankom XT10 (Ankom Technology Corp., Macedon, NY, USA), according to the American Oil Chemists' Society (AOCS) Official Procedure Am 5-04 [31]. Protein content was estimated by Kjeldahl method (N × 6.25). Dietary fiber contents (total (TDF), insoluble (IDF) and soluble (SDF) fiber) were determined with the dietary fiber assay kit TDF-100A (Sigma Aldrich, St. Louis, Missouri, U.S.A.) according to the AOAC procedures [32]. The total content of carbohydrate was calculated by difference. Protein and dietary fiber were also measured in the mushroom flours.

Na content was determined from the ashes of 5 g-aliquots of each sample dissolved in 10 mL of 1 M HNO_3, and analyzed by inductively coupled plasma (ICP)-optical emission spectroscopy (Thermo-Fisher, Cambridge, UK) equipped with a radio frequency source of 27.12 MHz, a peristaltic pump, a spraying chamber and a concentric spray nebuliser, and controlled by the ICP software. Standard solutions (50, 100, 150, 200 mg/L) were prepared from a stock solution of Na (1000 mg/L; SCP-SCIENCE, Courtaboeuf, France) in 4% HNO_3 (*v/v*). The results were expressed as milligrams per 100 g sausage.

2.4. Amino Acid Profile

Amino acid composition (g/100 g of sample) of frankfurters was determined using the methodology described by Marti-Quijal et al. [33] by derivatization of amino acids with 6-Aminoquinolyl-N-hydroxysuccinimidyl carbamate (Waters AccQ-Fluor reagent kit) and reversed-phase high-performance liquid chromatography analysis (RP-HPLC) (Waters 2695 Separations Module + Waters 2475 Multi Fluorescence Detector + WatersAccQ-Tag amino acids analysis column). Empower 2 TM advanced software (Waters, Milford, MA, USA) was used to control system operation and results management. Amino acids were identified by retention time using an amino acid standard (Amino Acid Standard H, Thermo, Rockford, IL, USA).

2.5. Lipid Oxidation Analysis

The thiobarbituric acid-reactive substances (TBARs) assay was carried to evaluate lipid stability according to the methodology of Vyncke [34]. For this purpose, 2 g of sample and 10 mL of trichloroacetic acid (5%) were homogenised using an Ultra-Turrax (IKA T25 basic, Staufen, Germany) for 2 min. The homogenate was kept at −10 °C for 10 min and centrifuged at 2360 g for 10 min. The supernatant was filtered through a Whatman No. 1 (Sigma Aldrich, St. Louis, MO, USA) filter paper. The extract (5 mL) was mixed with a 0.02 M thiobarbituric acid solution (5 mL) and incubated in a water bath at 97 °C for 40 min. The absorbance was measured at 532 nm. For quantification a standard curve of malondialdehyde (MDA) was designed, and the results were expressed as milligrams of MDA per kilogram of sample.

2.6. pH and Microbial Analysis

The pH from three frankfurters of each sample was measured with a digital pH-meter (HI 99163, Hanna Instruments, Eibar, Spain) equipped with a glass probe for penetration.

Total viable counts (TVC) and lactic acid bacteria were determined using TEMPO system (TEMPO Filler, TEMPO Reader, BioMérieuxs, Marcy l'Etoile, France), based on the most likely number technology. Both microorganisms were incubated at 30 °C for 24 and 48 h respectively (detection limit of 0.3 CFU/g of sample). *Pseudomonas* spp. counts were enumerated on Pseudomonas agar base with selective supplement for this microbial group (CFC) (Merck, Darmstadt, Germany) after incubation at 25 °C for 48 h. Psychrotrophic aerobic bacteria were enumerated on plate count agar (PCA; Oxoid, Unipath Ltd., Basingstoke, UK) following incubation at 7 °C for 10 days. Detection limits were 2 log CFU/g for pseudomonads and psychrotrophic bacteria. The microbial results were expressed as log CFU/g.

2.7. Color and Texture Profile

CIELAB parameters (L*: lightness; a*: redness and b*: yellowness) were measured in slices of 2 cm thickness using a portable colorimeter (Konica Minolta CM-600d, Osaka, Japan) under D65 illuminant and 10° observer with an 8 mm aperture.

A texturometer (TA-XTplus, Stable Micro Systems, Surrey, UK) equipped with Texture Exponent 32 software (version 1.0.0.68) was used to measure the texture profile analysis (TPA). The samples were cut in pieces of 2.5 cm height × 2 cm diameter and hardness (N), springiness (mm), chewiness (N/mm), gumminess (N) and cohesiveness (mm/mm) were determined (Bourne, 1978). Textural parameters were measured by compressing to 50% with a double compression cycle test using an aluminum cylinder probe P50 (50 mm diameter) at speed of 20 mm/s and a distance of 30 mm with a 50 Kg load cell. Three slices of each sample were measured.

2.8. Sensory Analysis

Sensory evaluation of frankfurters- type sausage was conducted by twenty trained panelists selected from the members of Meat Technology Center of Galicia, being the participants trained according ISO regulations [35] with the attributes and the scale to evaluate the color, discoloration at surface and odor of raw sausages stored for 0, 30, 60 and 90 days, using a 5-point hedonic scale, ranging from 1 = "excellent" to 5 = "not acceptable" [36]. Also, odor and taste of 1 cm-slices cooked at 180 °C until reach an internal temperature of 72 °C, were evaluated at day 0. Bread without salt and water was used to clean the palate between samples.

2.9. Statistical Analysis

Analysis of variance (ANOVA) was performed with Statgraphics Centurion XVI version 16.1.03 (StatPoint Technologies, Inc., Warrenton, VA, USA). Tukey's test was used to compare the mean values at a significance level of $p < 0.05$.

3. Results and Discussion

3.1. Chemical Composition

The results of chemical composition of different treatments are shown in Table 2. The addition of mushroom flour resulted in a significant ($p < 0.05$) moisture increase (64.34–66.48%) compared with the control (61.05%). The higher concentration of flour resulted in the highest moisture values in LF-Ab and LF-Po. These moisture increments could be attributed to the higher water amount added to compensate the fat reduction in formulation and the presence of mushroom flour, which is rich in dietary fiber, particularly β-glucans [37]. Contents of total dietary fiber of the mushroom flours Ab and Po were 22.82 ± 0.42% and 43.58 ± 1.33% in dry weight, respectively. Several works have pointed out that the porous and hydrophilic capacity of dietary fiber contributes to water holding properties increasing moisture [38,39]. The increased in moisture has been reported in other fat replacers like konjac gel [40], or pineapple dietary fibers [12].

Table 2. Proximate composition (%) and content sodium (mg/100 g) of frankfurters type sausage elaborated with flour of edible mushroom (*Agaricus bisporus* and *Pleurotus ostreatus*).

	Control	MF-Ab	MF-Po	LF-Ab	LF-Po	MF-PoAb
Moisture	61.05 ± 0.17 [a]	64.34 ± 0.13 [b]	65.91 ± 0.21 [d]	66.34 ± 0.50 [de]	66.48 ± 0.30 [e]	65.37 ± 0.55 [c]
Fat	19.16 ± 0.42 [f]	16.28 ± 0.33 [e]	15.18 ± 0.29 [d]	12.99 ± 0.30 [b]	11.79 ± 0.06 [a]	14.04 ± 0.47[c]
Protein	14.39 ± 0.14 [b]	14.70 ± 0.27 [b]	13.62 ± 0.22 [a]	15.41 ± 0.54 [c]	14.59 ± 0.33 [b]	14.29 ± 0.24 [b]
Ash	2.08 ± 0.05 [d]	1.30 ± 0.04 [a]	1.36 ± 0.06 [ab]	1.36 ± 0.08 [ab]	1.81 ± 0.03 [c]	1.41 ± 0.05 [b]
Carbohydrates	3.32 ± 0.25 [a]	3.37 ± 0.39 [a]	3.93 ± 0.31 [b]	3.90 ± 0.19 [b]	5.33 ± 0.04 [d]	4.88 ± 0.21 [c]
Dietary fiber	0.05 ± 0.09 [a]	0.57 ± 0.06 [b]	1.09 ± 0.11 [c]	1.14 ± 0.09 [c]	2.18 ± 0.25 [e]	1.58 ± 0.17 [d]
Na (mg/100g)	686.30 ± 37.10 [c]	338.69 ± 40.53 [a]	336.46 ± 17.97 [a]	345.22 ± 30.47 [a]	405.19 ± 12.53 [b]	347.48 ± 4.07 [a]

Results are expressed as mean value ± standard deviation. [a–f]: Different letters in each batch indicate significant differences ($p < 0.05$).

As expected, batches formulated with lower content of fat and salt resulted in significant ($p < 0.05$) lower concentrations for these parameters due to the modification in the formulations. Fat was reduced from 19.16 ± 0.42 to 16.28 ± 0.33, 15.18 ± 0.29 and 14.04 ± 0.47 in medium fat sausages MF-Ab, MF-Po and MF-PoAb, respectively. In LF-Ab (12.99 ± 0.30) and LF-Po (11.79 ± 0.06) was reached the highest reduction of fat. Na contents of flour added samples were approximately half of the control samples. The same behavior was observed in the ash content, quite related with the sodium chloride content reduction in the formulations [36]. The reduction of fat in meat emulsions can provoke changes in emulsion stability parameters, such as fat and water losses during cooking. Therefore, the meat industry has adopted new trends to improve the texture and water holding capacity, substituting animal fat by including the use of non-meat ingredients, such as inulin or β-glucan, considering the fiber's ability to retain fat and water [15]. However, when Ab and Po performance are compared, Po was less effective in fat retention, probably related to different protein content of flours, 28.63 ± 0.10% in Ab flour and 16.04 ± 0.22% in Po flour, expressed in dry basis. When the protein content of frankfurters was analyzed, only sausages with 5% Ab mushroom presented a significant higher protein content ($p < 0.05$) despite the higher protein content of Ab flour in agreement with the reported by Reis et al. [41] and Cheung [42]. The expected increase in this parameter because of the mushroom addition was overshadowed by the moisture augmentation reached in mushroom flour added products and the reduction of 50% of sodium caseinate in formulations. When the protein content is reported in dry weight (DW) all mushroom formulations presented significant higher contents (over 40% DW) than the 36.94 DW from control samples, being the LF-Ab, LF-Po the samples with the major values 45.77 g/100 g and 43.52 g/100 g DW, respectively.

The addition of these flours as a source of fiber in meat products has to be highlighted, since fiber content is usually absent in these products. Several works have included different vegetables, legumes or fiber sources to increase the fiber content of these products [11,12]. The batches with addition of the Ab and Po flour showed higher fibre content in a range of 0.57 and 2.18 g/100 g of sausage in comparison with the control (0.05 g/100 g sausage). LF-Po presented the higher value of dietary fibre with 6.51 g/100 g sausage (DW) since the Po flour presented the highest content in dietary fiber. The dietary fiber in mushrooms comes from non-digestible carbohydrates mainly chitin, glucans, cellulose and hemicelluloses like mannans, xylans and galactans [18,42]. With the combination of both mushroom flours in MF-PoAb the reduction of fat is accompanied with an interesting fiber content of 1.58 ± 0.17 being a promising alternative for a healthier sausage keeping the protein content with half of the sodium caseinate added.

3.2. Amino Acid Profile

When the amino acid profile of the treatments is analyzed (Table 3) no significant changes were observed between batches ($p > 0.05$). Some studies have been focused on the addition of non-meat protein sources like legumes or even algae in order to improve the protein profile in meat products [10,33]. In this case the addition of mushroom flour did not modify the amino acid profile keeping the ratio between essential and no essential amino acids in 0.93–0.94. The predominant amino acids in formulations were glutamic acid (ranging 2.58–2.84 g/100 g), aspartic acid (1.44–1.66 g/100 g), lysine (1.38–1.54 g/100 g) and leucine (1.31–1.43 g/100 g). These amino acids with alanine and arginine are also the most important in dried edible mushrooms [18], being glutamic and aspartic acids strongly related to umami taste [17]. However, the concentrations of flours added did not change the amino acid profile although sensorially mushroom taste was noticed. Related to essential amino acids lysine, leucine, arginine and valine were the predominant essential amino acids. The edible mushroom has been reported to contain all nine essential amino acids required for human intake [17,43], even though the protein profile of mushrooms depends not only on the specie but also on the size, composition of the substrate and harvest time [18]. In the case of *A. bisporus* and *P. ostreatus* leucine has been reported to be the limiting amino acid but with a protein quality in terms of digestibility and essential amino acid composition comparable to casein, eggs, and soy [17,37]. In this work, 50% of caseinate was substituted by mushroom flour in the formulation, but considering that non significant differences ($p > 0.05$) were found regarding the essential amino acid profiles, mushroom flours could be even suggested to partially replace meat content or to be considered as meat substitute.

Table 3. Amino acid profile (g/100 g sample) of protein from frankfurters with addition of edible mushroom flours.

	Control	MF-Ab	MF-Po	LF-Ab	LF-Po	MF-PoAb
OH-Proline	0.21 ± 0.01	0.18 ± 0.01	0.15 ± 0.01	0.18 ± 0.01	0.18 ± 0.06	0.19 ± 0.02
Aspactic acid	1.44 ± 0.20	1.52 ± 0.03	1.49 ± 0.01	1.66 ± 0.20	1.44 ± 0.29	1.62 ± 0.19
Serine	0.65 ± 0.07	0.64 ± 0.03	0.64 ± 0.01	0.72 ± 0.06	0.65 ± 0.11	0.73 ± 0.03
Glutamic acid	2.67 ± 0.37	2.65 ± 0.01	2.59 ± 0.01	2.81 ± 0.29	2.58 ± 0.55	2.84 ± 0.26
Glycine	0.95 ± 0.06	0.85 ± 0.01	0.92 ± 0.01	0.95 ± 0.05	1.00 ± 0.21	1.01 ± 0.01
Alanine	0.88 ± 0.10	0.88 ± 0.01	0.91 ± 0.03	0.93 ± 0.10	0.90 ± 0.23	0.95 ± 0.08
Cysteine	0.10 ± 0.01	0.09 ± 0.01	0.09 ± 0.02	0.11 ± 0.01	0.11 ± 0.02	0.12 ± 0.01
* Valine	0.91 ± 0.14	0.89 ± 0.04	0.90 ± 0.06	0.96 ± 0.12	0.92 ± 0.24	0.98 ± 0.04
Methionine	0.40 ± 0.05	0.40 ± 0.01	0.38 ± 0.01	0.41 ± 0.10	0.42 ± 0.09	0.48 ± 0.06
* Isoleucine	0.81 ± 0.15	0.80 ± 0.05	0.79 ± 0.04	0.85 ± 0.12	0.79 ± 0.19	0.87 ± 0.04
* Leucine	1.35 ± 0.24	1.33 ± 0.09	1.31 ± 0.08	1.41 ± 0.19	1.31 ± 0.32	1.43 ± 0.06
Tyrosine	0.40 ± 0.05	0.40 ± 0.02	0.38 ± 0.04	0.46 ± 0.04	0.42 ± 0.07	0.44 ± 0.03
* Phenylalanine	0.69 ± 0.10	0.67 ± 0.06	0.67 ± 0.04	0.73 ± 0.08	0.69 ± 0.13	0.73 ± 0.01
* Hystidine	0.54 ± 0.03	0.51 ± 0.01	0.52 ± 0.01	0.57 ± 0.06	0.56 ± 0.11	0.58 ± 0.01

Table 3. *Cont.*

	Control	MF-Ab	MF-Po	LF-Ab	LF-Po	MF-PoAb
* Lysine	1.45 ± 0.20	1.50 ± 0.04	1.46 ± 0.01	1.54 ± 0.12	1.38 ± 0.28	1.53 ± 0.22
* Arginine	0.93 ± 0.09	0.92 ± 0.03	0.90 ± 0.01	0.97 ± 0.05	0.90 ± 0.14	0.97 ± 0.12
Proline	1.05 ± 0.17	0.97 ± 0.04	0.96 ± 0.07	1.02 ± 0.13	0.94 ± 0.34	1.05 ± 0.02
Taurine	<0.01	<0.01	<0.01	<0.01	<0.01	<0.01
* Threonine	0.73 ± 0.08	0.73 ± 0.03	0.71 ± 0.02	0.78 ± 0.04	0.72 ± 0.12	0.80 ± 0.07
Total	16.16 ± 2.09	15.94 ± 0.39	15.79 ± 0.29	17.07 ± 1.77	15.93 ± 3.51	17.35 ± 1.06
Essential (E)	7.81 ± 1.09	7.74 ± 0.27	7.64 ± 0.20	8.22 ± 0.88	7.70 ± 1.63	8.38 ± 0.51
Non-essential (Ne)	8.35 ± 1.00	8.19 ± 0.12	8.14 ± 0.09	8.85 ± 0.89	8.23 ± 1.88	8.97 ± 0.55
E/Ne	0.93 ± 0.02	0.94 ± 0.02	0.94 ± 0.01	0.93 ± 0.01	0.94 ± 0.02	0.93 ± 0.01

*: Essential amino acid.

3.3. Lipid Oxidation Analysis

The oxidative process was evaluated during storage time at 0, 30, 60 and 90 days, and the results are shown in Figure 1. The analysis of variance indicated that TBA values were significantly ($p < 0.05$) affected by the concentration of the edible mushroom flours added and the storage time. Initially, the scores obtained in samples with mushroom flours and less fat were significantly higher ($p < 0.05$) ranging in 0.34–0.72 mg MDA/Kg than the found in control samples with 0.12 mg MDA/Kg. This behaviour remained during cold storage. However, most treatments with Ab and Po flours presented TBA values below the acceptable limit (<1.0 mg MDA/kg) [44]. Mushrooms have been described to possess antioxidant activity, mainly because of the phenolic compounds, although their antioxidant properties depend on the species of mushroom, and the growing, harvest and processing conditions [45,46]. However, the antioxidant effect mushroom flours was not appreciated in this study and the higher concentration of flour (LF-Ab and LF-Po) led to higher initial TBARs, when a reduction of TBARs should be expected, since in these samples a reduction of 50% of fat was applied.

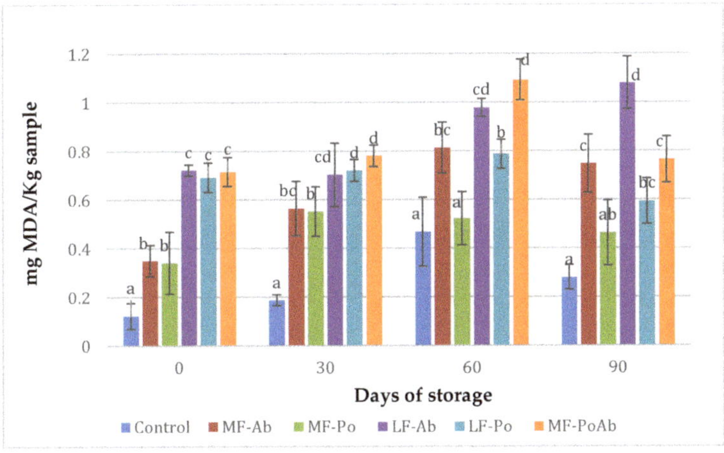

Figure 1. Results of TBARS (mg MDA/Kg) of sausages added with Ab and Po flours. (a–d): Mean values (corresponding to the same day) not followed by a common letter differ significantly ($p < 0.05$).

The limited antioxidant activity of mushroom flour and high TBARS found initially in samples with Ab and Po flour could be due to the drying conditions applied to the mushrooms during the obtaining of the flour, which could have promoted browning reaction and protein degradation products, participating in the formation of TBA color complexes and overestimating MDA values, as it has been reported by Papastegriadis et al. [47] in products like dry nuts. Besides, a significant increase was observed during storage for all treatments but after 90 days of cold storage, TBARS values significantly decreased ($p > 0.05$). This behavior was also noticed by Fernandes et al. [48] and attributed to instability or transitory nature of some secondary products from lipid peroxidation like malondialdehyde (MDA). But also, the higher values of TBARs of Ab samples especially from day 60 comparing to Po samples could be attributable to more browning reaction products present in Ab samples reacting with TBA since Ab flour have a darker color.

3.4. pH and Microbial Results

The inclusion of mushroom flour in the sausage led to a significant increase of pH over 6.0 comparing to 5.94 ± 0.02 Log CFU/g of control samples and the pH values kept during the cold storage. The pH of samples was in the range (5.94–6.11) similar to the pH reported for frankfurters with different levels of shiitake [8], and other cooked sausages with extracts or alternative ingredients added [48]. Microbiologically, the cooking process and post-packing pasteurization eliminated the vegetative microorganisms, so lactic acid bacteria, pseudomonads and psychrotrophic bacteria remained under detection limits. However, counts between 4.52 ± 0.14–6.12 ± 0.08 CFU/g were found in total viable counts for flour mushroom added-frankfurters, much higher than the 1.68 CFU/g reported for control samples, and remained stable during the cold storage (Figure 2). These high levels of microorganisms can be attributed to spore forming bacteria naturally present in agrifoods in contact with soil like vegetables and mushrooms, which survived to the thermal treatment. Ab flour sausages presented significant higher counts ($p < 0.05$) than Po samples, possibly related to the conditioning process, which involved a blanching process, spreading the spore-forming contamination. However, in all cases the high counts are not considered a risk as long as refrigeration temperatures are maintained.

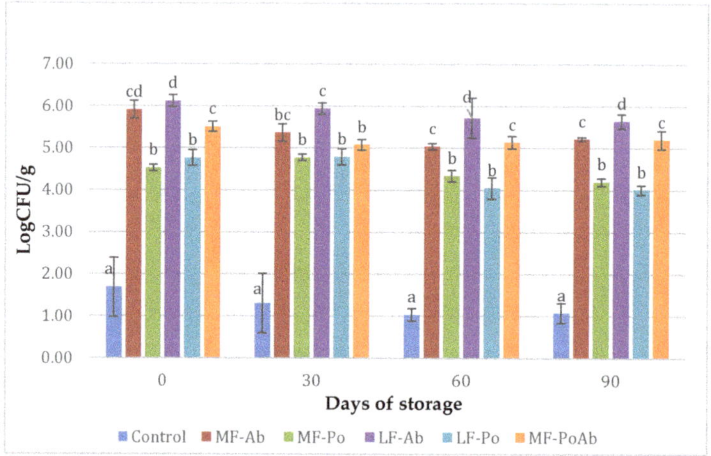

Figure 2. Evolution of microbial counts (Log CFU/g) of sausages during storage with Ab and Po flours. (a–d): Mean values (corresponding to the same day) not followed by a common letter differ significantly ($p < 0.05$).

3.5. Color and Texture Profile Analysis

The color parameters of frankfurters during storage are shown in Figure 3. The addition of mushroom flour significantly reduced ($p < 0.05$) the lightness by the fat reduction procedure, although Po flour decreased the L* parameter to a lesser extend (LF-Po: 60.63 ± 0.35–MF-Po: 63.90 ± 0.48) than Ab flour (LF-Ab: 57.57 ± 4.31–MF-Ab: 58.57 ± 0.64). Yellowness (b*) of the frankfurters increased in flour added frankfurters, especially in samples with Po. Ab samples showed significant lower a* values ($p < 0.05$) comparing with control samples, while addition of Po flour significantly ($p < 0.05$) increased this parameter. So, Ab flour contributed to significantly ($p < 0.05$) reduce the L* and a* values comparing to the control giving a darker color to the frankfurters as can be seen in Figure 4. Visually, the characteristic pink color of the control samples due to nitrosomyoglobine formation was slightly modified into brownish colors. The addition of edible mushroom flours instead of animal fat did not reproduce the color effect of pork backfat on the treated samples, especially in the Ab samples. Fat contributes in emulsions to lighter meat products and usually natural and artificial colorants are added to keep the pink color. The replacement of pork fat from frankfurters or finely comminuted sausages by oleogels [49] have also resulted in lighter and less red meat products. From the brown or greenish color by the incorporation of algae [10,50,51] or chia [11] to the to orange tones by addition of lycopene or carotenoids [52], the change of color will depend not only on the ingredient source, but also on the concentration added [12]. Finally the presence of mushroom flours was not negatively scored in sensory analysis.

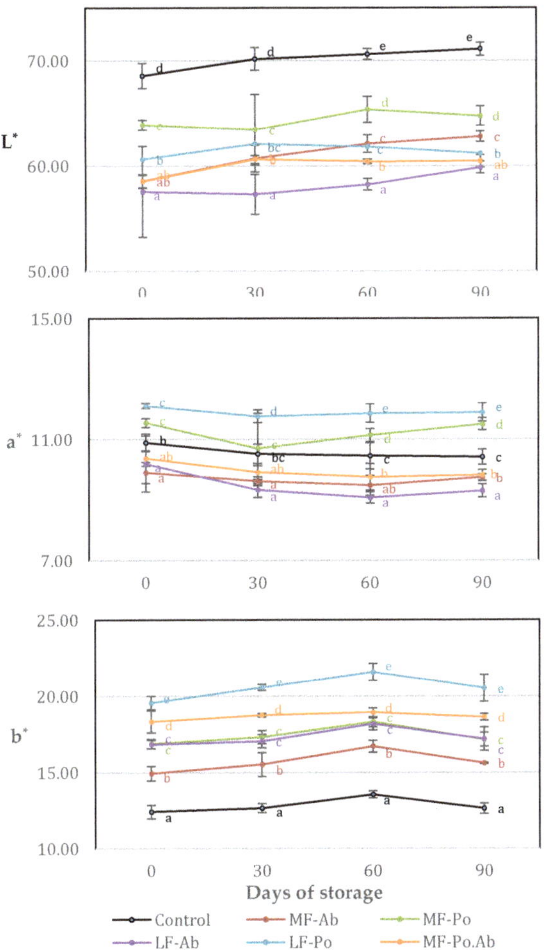

Figure 3. Evolution of color parameters (**L***, **a*** and **b***) during the storage of cooked sausages with edible mushroom flours (a–e): Mean values (corresponding to the same day) not followed by a common letter differ significantly (*p* < 0.05).

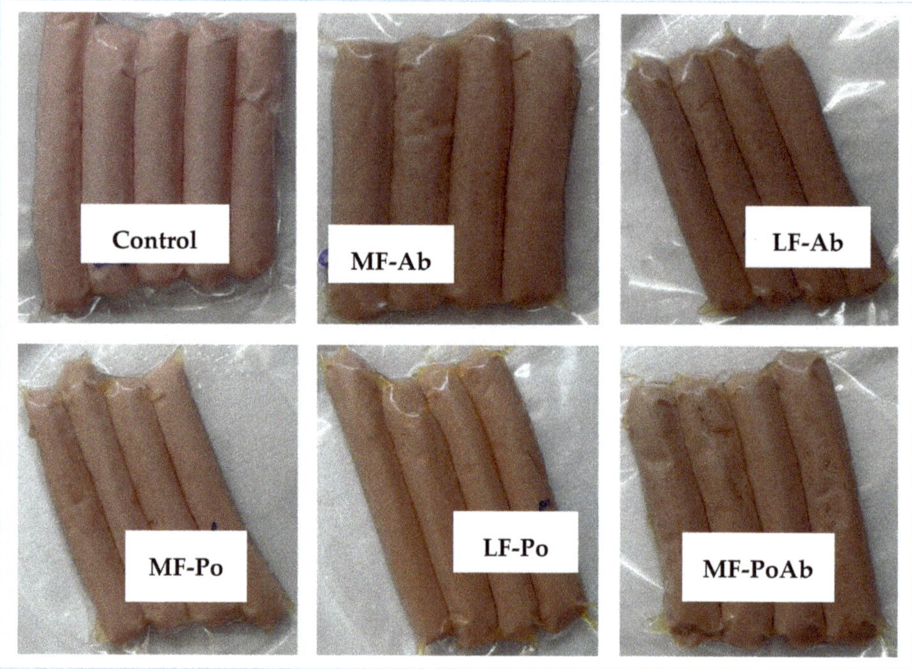

Figure 4. Visual appearance of frankfurter with different formulations.

Cold storage marginally affected to color parameters as it has been reported with other fat substitutes in frankfurters like chia or vegetal oils in konjac matrix [11,40]. Although natural pigments use to presenting less color stability [52], color frankfurters with mushroom flour remained mostly stable although a slight increase was observed in lightness probably due to the oxidation of fat. No typical discoloration was observed because of the oxidation of myoglobin pigments during storage [11,12].

The addition of 2.5% and 5% of edible mushroom flours also significantly affected ($p < 0.05$) the texture parameters evaluated in frankfurter with partial decrease of fat (30% and 50%) and salt (Table 4). All mushroom added-samples showed significant ($p < 0.05$) lower hardness, springiness, cohesiveness, gumminess and chewiness values comparing to control samples after elaboration. Samples with Po flour (2.5 and 5%; MF-Po and LF-Po), even in combination with Ab flour (MF-PoAb), gave the lowest values in the textural parameters resulting in softer frankfurters. In the emulsions fat is dispersed in small drops in a continuous phase formed by water, proteins and additives. When the fat is reduced the emulsion can loose stability and affect the texture. In general, when the reduction of fat is compensated by increasing protein, textures tend to be harder. However, when fat content is reduced by increasing the proportion of water, keeping the amount of protein, the structure of low-fat systems becomes softer [40]. In this case the replacement of fat and addition of mushroom flour led to a higher water contents in formulation and protein and fibre from mushrooms could help to bound water reducing hardness, and the rest of textural parameters with concentration of flour added. However, differences found according to the origin of mushroom flour, could be related to the different protein content of mushroom flours, since Ab flour (with a higher protein content) reduced to a lesser extend the textural parameters than Po flour (with around 16% of protein but a higher fiber content, over 40%).

Proteins from different sources have been added to meat products like frankfurters to stabilize and compensate the reduction of meat or fat. Stephan et al. [53] reported suitable properties of mycelia from *Pleurotus sapidus* as meat substitute in vegan boiled sausages comparable to the use of other protein concentrates (soy, pea and sunflower), but lower textural parameters than the original german boiled sausage were observed. In this case, the addition of Ab mushroom flour better helped to stabilize the emulsion, even if other protein source like sodium caseinate was reduced in the formulation.

Table 4. Texture parameters of frankfurters with edible mushroom flours.

Texture Parameter	Days of Storage			
	0	30	60	90
Hardness (N)				
Control	20.56 ± 1.01 [d,X]	25.12 ± 2.79 [c,Y]	24.18 ± 0.06 [d,XY]	27.84 ± 3.90 [c,Y]
MF-Ab	15.33 ± 2.63 [c,X]	15.15 ± 0.94 [ab,X]	16.33 ± 1.83 [c,X]	17.46 ± 1.46 [b,X]
MF-Po	13.53 ± 2.26 [bc,X]	16.83 ± 1.69 [b,Y]	16.40 ± 2.15 [c,XY]	14.75 ± 1.98 [ab,XY]
LF-Ab	12.07 ± 1.25 [ab,X]	16.42 ± 1.17 [b,Y]	13.67 ± 1.83 [ab,X]	16.09 ± 1.05 [b,Y]
LF-Po	10.84 ± 0.44 [a,X]	13.41 ± 1.62 [a,Y]	12.52 ± 1.40 [a,XY]	13.09 ± 0.69 [ab,Y]
MF-PoAb	11.50 ± 0.95 [ab,X]	13.02 ± 0.66 [a,X]	16.08 ± 1.68 [bc,Y]	14.93 ± 1.25 [ab,Y]
Springiness (mm)				
Control	0.78 ± 0.03 [c,X]	0.82 ± 0.01 [c,Y]	0.81 ± 0.01 [e,Y]	0.81 ± 0.01 [c,Y]
MF-Ab	0.70 ± 0.04 [b,X]	0.73 ± 0.04 [b,XY]	0.75 ± 0.02 [d,Y]	0.74 ± 0.03 [bc,XY]
MF-Po	0.59 ± 0.04 [a,X]	0.65 ± 0.05 [a,X]	0.64 ± 0.05 [ab,X]	0.63 ± 0.03 [a,X]
LF-A	0.66 ± 0.02 [b,X]	0.74 ± 0.02 [b,Y]	0.71 ± 0.03 [cd,Y]	0.75 ± 0.02 [c,Y]
LF-Po	0.55 ± 0.02 [a,X]	0.68 ± 0.08 [ab,Y]	0.62 ± 0.04 [a,XY]	0.69 ± 0.05 [b,Y]
MF-PoAb	0.59 ± 0.03 [a,X]	0.65 ± 0.05 [a,XY]	0.69 ± 0.03 [bc,Y]	0.70 ± 0.05 [bc,Y]
Cohesiveness (mm/mm)				
Control	0.38 ± 0.01 [c,X]	0.44 ± 0.06 [c,Y]	0.42 ± 0.02 [d,XY]	0.39 ± 0.01 [e,XY]
MF-Ab	0.34 ± 0.01 [bc,X]	0.34 ± 0.01 [ab,X]	0.36 ± 0.01 [c,X]	0.36 ± 0.02 [cd,X]
MF-Po	0.29 ± 0.03 [a,X]	0.31 ± 0.03 [a,X]	0.31 ± 0.01 [a,X]	0.29 ± 0.02 [a,X]
LF-Ab	0.34 ± 0.01 [bc,X]	0.37 ± 0.01 [b,YZ]	0.34 ± 0.02 [bc,XY]	0.38 ± 0.02 [de,Z]
LF-Po	0.31 ± 0.02 [ab,X]	0.34 ± 0.02 [ab,Y]	0.33 ± 0.01 [ab,XY]	0.33 ± 0.02 [b,XY]
MF-PoAb	0.31 ± 0.03 [ab,X]	0.32 ± 0.01 [a,X]	0.33 ± 0.02 [ab,X]	0.34 ± 0.01 [bc,X]
Gumminess (N)				
Control	7.73 ± 0.32 [c,X]	10.79 ± 1.44 [c,Y]	10.06 ± 0.30 [d,Y]	10.81 ± 1.53 [d,Y]
MF-Ab	5.24 ± 1.27 [b,X]	5.14 ± 0.45 [ab,X]	5.86 ± 0.78 [c,X]	6.29 ± 0.63 [c,X]
MF-Po	3.99 ± 1.03 [a,X]	5.22 ± 0.74 [ab,Y]	5.11 ± 0.52 [abc,XY]	4.34 ± 0.80 [a,XY]
LF-Ab	4.10 ± 0.53 [ab,X]	6.00 ± 0.27 [b,Y]	4.73 ± 0.84 [ab,X]	6.04 ± 0.46 [bc,Y]
LF-Po	3.39 ± 0.26 [a,X]	4.59 ± 0.72 [a,Y]	4.13 ± 0.47 [a,Y]	4.30 ± 0.20 [a,Y]
MF-PoAb	3.66 ± 0.71 [a,X]	4.18 ± 0.45 [a,XY]	5.32 ± 0.88 [bc,Z]	5.05 ± 0.34 [ab,YZ]
Chewiness (N·mm)				
Control	6.01 ± 0.46 [c,X]	8.80 ± 1.09 [d,Y]	8.21 ± 0.13 [d,Y]	8.82 ± 1.17 [c,Y]
MF-Ab	3.67 ± 0.98 [b,X]	3.77 ± 0.29 [bc,X]	4.41 ± 0.75 [c,X]	4.65 ± 0.61 [b,X]
MF-Po	2.37 ± 0.75 [a,X]	3.41 ± 0.75 [ab,Y]	3.31 ± 0.59 [ab,XY]	2.71 ± 0.42 [a,XY]
LF-Ab	2.70 ± 0.34 [a,X]	4.46 ± 0.31 [c,Y]	3.40 ± 0.73 [ab,X]	4.52 ± 0.48 [b,Y]
LF-Po	1.87 ± 0.20 [a,X]	3.15 ± 0.82 [ab,Y]	2.58 ± 0.41 [a,XY]	2.98 ± 0.33 [a,Y]
MF-PoAb	2.17 ± 0.51 [a,X]	2.74 ± 0.45 [a,XY]	3.70 ± 0.80 [bc,Z]	3.56 ± 0.39 [a,YZ]

Results are expressed as mean value ± standard deviation. (X–Z): Means in the same row not followed by a common letter are significantly different ($p < 0.05$). (a–e): Mean values in the same column (for each texture parameter) not followed by a common letter are significantly different ($p < 0.05$).

In general, softer textures have been obtained when different vegetal, cereal or legume, algae or fiber sources have been included in meat formulations with or without replacement of fat or other ingredients. Alvarez et al. [15] found a significant decrease of hardness in frankfurters by addition of rice bran and walnut. Choi et al. [44] reported a significant decrease of hardness in reduced- fat frankfurters with vegetable oils and rice bran fiber, and Cofrades et al. [54] observed that increasing amounts of walnut extracts reduced shear force and elongation values, indicating the formation of softer and less cohesive meat structures. Addition of 1% of pineapple fiber in partial replacement of fat in sausages significantly reduced hardness, chewiness and gumminess but less differences were appreciated in cohesiveness and springiness [12]. The partial replacement of fat and salt by addition of 1% of different seaweeds resulted also in lower hardness and chewiness parameters although the behavior on adhesiveness and springiness was seaweed-species related [50]. The addition of concentration of 0.8–1.2% of shiitake (*Lentinus edodes*) powder in frankfurter also reduced hardness, but increased cohesiveness [8].

In general, texture parameters were also affected ($p < 0.05$) by cold storage increasing all the parameters, but especially hardness in control samples comparing with mushroom added sausages. An increase of hardness during storage has been reported in frankfurters as a consequence to the increase of purge loss [40].

3.6. Sensory Evaluation

In the case of sensory attributes evaluated in samples before cooking (color, decoloration surface), no significant differences ($p > 0.05$) were detected between treatments with color values between good and acceptable (2.17–2.83) and no discoloration was observed (Figure 5). The evaluation of odor indicated that samples with Ab (MF-Ab, LF-Ab, MF-PoAb) and LF-Po presented a stronger mushroom odor, with scores over acceptable level, while MF-Po did not significantly ($p > 0.05$) differ from control. When the samples were cooked the reduction of fat and salt and the inclusion of Ab and Po flours resulted in lower sensorial values for flavor and taste but around acceptability level, without significant differences between flour added samples. In the case of flavor, the samples with 5% of flour (LF-Ab, LF-Po and MF-PoAb) scored over acceptable level (3.0) while control samples scored near good level (1.95 ± 0.76). In the taste parameter, only LF-Ab scored over acceptable level, while de rest of treatments were rated between good and acceptable levels. The characteristic flavor or umami in mushrooms is intense, especially in Ab samples [20], and the acceptability of the frankfurter will depend on how accustomed the consumer to this flavor. When aqueous extract from other edible mushroom, *Cantharella cibarius*, was added to frankfurter, it was not sensorially different from control samples [26]. In general, umami compounds enhance palatability in savory foods which allows to reduce salt content without a negatively perception in the consumer increasing also satiety [55]. But the incorporation of new ingredients into processed meats modifies sensory properties and can limit the consumer acceptability [10]. However, this untypical or unexpected flavor could be moderated by incorporating different seasonings, since no flavor additives were included in formulations here tested, and even control samples were scored as good, but not excellent, because of the plain flavor. Jimenez-Colmenero et al. [51] showed that the addition of konjac-seaweed flour to low fat and low salt frankfurters slightly reduced sensory panel values, due to an intense unfamiliar flavor. However, they did not consider this an obstacle, and suggested the reformulation using less strongly flavored seaweed and considering seasonings.

During cold storage, sensory parameters evaluated in uncooked samples, color, discoloration and odor did not significantly changed confirming the stability of the product during long periods of storage.

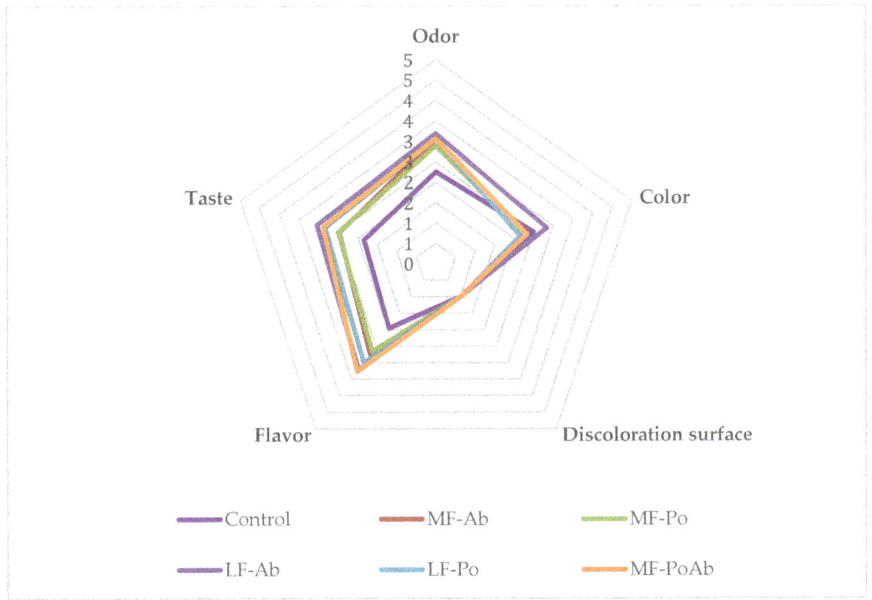

Figure 5. Sensory results of frankfurters evaluated at day 0.

4. Conclusions

The inclusion of Ab and Po flours in fat- and salt-reduced frankfurter sausages seems to a feasible strategy to enhance the nutritional profile these products, although the physicochemical, textural and sensory properties were affected by the mushroom and concentration source. The protein profile did not changed, but fiber contents were improved, which makes Ab and Po flour an interesting substitute for fat and salt and even for meat. Color was significantly affected, especially by the inclusion of Ab flour, resulting in darker products, while Po flour addition presented a higher impact in the texture with softer and less cohesive sausages. Nevertheless, all mushroom flours remained sensorially acceptable despite the strong umami flavor was perceived. During cold storage, samples remained quite unchanged for 90 days.

Author Contributions: Conceptualization, E.M.S., E.R.-V.; methodology, R.B., M.P. and J.A.R.; formal analysis, investigation, M.I.C.-G., E.M.S. and E.R.-V.; writing—original draft preparation, M.I.C.-G. and E.M.S.; writing—review and editing, J.M.L., M.P. and I.S.-O.; Funding acquisition: E.M.S. and J.M.L. All authors have read and agreed to the published version of the manuscript.

Funding: This research was funded by Empacadora Murgati and CONACyT—the Incentives for Research, Technological Development and Innovation Program (conv. 2016-232144) and SEP Program to Improve Educational Quality "Health care network through -strategies based on food and natural products" (P/PFCE-2018-13MSU0017T-04)". Authors are members of the Healthy Meat network, funded by CYTED (grant number 119RT0568).

Conflicts of Interest: The authors declare no conflict of interest. The funders had no role in the design of the study; in the collection, analyses, or interpretation of data; in the writing of the manuscript, or in the decision to publish the results.

References

1. Hathwar, S.C.; Rai, A.K.; Modi, V.K.; Narayan, B. Characteristics and consumer acceptance of healthier meat and meat product formulations: A review. *J. Food Sci. Technol.* **2012**, *49*, 653–664. [CrossRef]
2. Decker, E.A.; Park, Y. Healthier meat products as functional foods. *Meat Sci.* **2010**, *86*, 49–55. [CrossRef]

3. Ekmekcioglu, C.; Wallner, P.; Kundi, M.; Weisz, U.; Haas, W.; Hutter, H.-P. Read meat, diseases, and healthy alternatives: A critical review. *Crit. Rev. Food Sci. Nutr.* **2018**, *58*, 247–261. [CrossRef] [PubMed]

4. Beriain, M.J.; Gomez, I.; Ibañez, F.C.; Sarries, M.V.; Ordoñez, A.I. Improvement of the functional and healthy properties of meat products. In *Food Quality: Balancing Health and Disease*, 1st ed.; Holban, A.M., Grumezescu, A.M., Eds.; Academic Press: London, UK, 2018; Volume 13, pp. 1–74. ISBN 9780128114940.

5. Grasso, S.; Brunton, N.P.; Lyng, J.G.; Lalor, F.; Monahan, F.J. Healthy processed meat products—Regulatory, reformulation and consumer challenges. *Trends Food Sci. Technol.* **2014**, *39*, 4–17. [CrossRef]

6. Zhang, W.; Xiao, S.; Samaraweera, H.; Joo, E.; Ahn, D.U. Improving functional value of meat products. *Meat Sci.* **2010**, *86*, 15–31. [CrossRef] [PubMed]

7. Panagiotopoulou, E.; Moschakis, T.; Katsanidis, E. Sunflower oil organogels and organogel-in-water emulsions (part II): Implementation in frankfurter sausages. *LWT Food Sci. Technol.* **2016**, *73*, 351–356. [CrossRef]

8. Pil-Nam, S.; Park, K.M.; Kang, G.H.; Cho, S.H.; Park, B.Y.; Van-Ba, H. The impact of addition of shiitake on quality characteristics of frankfurter during refrigerated storage. *LWT Food Sci. Technol.* **2015**, *62*, 62–68. [CrossRef]

9. Delgado-Pando, G.; Cofrades, S.; Ruiz-Capillas, C.; Teresa Solas, M.; Jiménez-Colmenero, F. Healthier lipid combination oil-in-water emulsions prepared with various protein systems: An approach for development of functional meat products. *Eur. J. Lipid Sci. Technol.* **2010**, *112*, 791–801. [CrossRef]

10. Cofrades, S.; Benedí, J.; Garcimartin, A.; Sánchez-Muniz, F.J.; Jimenez-Colmenero, F. A comprehensive approach to formulation of seaweed-enriched meat products: From technological development to assessment of healthy properties. *Food Res. Int.* **2017**, *99*, 1084–1094. [CrossRef]

11. Pintado, T.; Herrero, A.M.; Jiménez-Colmenero, F.; Ruiz-Capillas, C. Strategies for incorporation of chia (*Salvia hispanica* L.) in frankfurters as a health-promoting ingredient. *Meat Sci.* **2016**, *114*, 75–84. [CrossRef]

12. Henning, S.S.C.; Tshalibe, P.; Hoffman, L.C. Physico-chemical properties of reduced-fat beef species sausage with pork back fat replaced by pineapple dietary fibres and water. *LWT Food Sci. Technol.* **2016**, *74*, 92–98. [CrossRef]

13. Choi, Y.S.; Kim, H.W.; Hwang, K.E.; Song, D.H.; Jeong, T.J.; Kim, Y.B.; Kim, C.J. Effects of fat levels and rice bran fiber on the chemical, textural, and sensory properties of frankfurters. *Food Sci. Biotechnol.* **2015**, *24*, 489–495. [CrossRef]

14. Özvural, E.B.; Vural, H. Which is the best grape seed additive for frankfurters: Extract, oil or flour? *J. Sci. Food Agric.* **2014**, *94*, 792–797. [CrossRef]

15. Alvarez, D.; Xiong, Y.L.; Castillo, M.; Payne, F.A.; Garrido, M.D. Textural and viscoelastic properties of pork frankfurters containing canola-olive oils, rice bran, and walnut. *Meat Sci.* **2012**, *92*, 8–15. [CrossRef]

16. Delgado-Pando, G.; Cofrades, S.; Ruiz-Capillas, C.; Solas, M.T.; Triki, M.; Jiménez-Colmenero, F. Low-fat frankfurters formulated with a healthier lipid combination as functional ingredient: Microstructure, lipid oxidation, nitrite content, microbiological changes and biogenic amine formation. *Meat Sci.* **2011**, *89*, 65–71. [CrossRef]

17. Bach, F.; Helm, C.V.; Bellettini, M.B.; Maciel, G.M.; Haminiuk, C.W.I. Edible mushrooms: A potential source of essential amino acids, glucans and minerals. *Int. J. Food Sci. Technol.* **2017**, *52*, 2382–2392. [CrossRef]

18. Deepalakshmi, K.; Mirunalini, S. *Pleurotus ostreatus*: An oyster mushroom with nutritional and medicinal properties. *J. Biochem. Technol.* **2014**, *5*, 718–726.

19. Dermiki, M.; Phanphensophon, N.; Mottram, D.S.; Methven, L. Contributions of non-volatile and volatile compounds to the umami taste and overall flavour of shiitake mushroom extracts and their application as flavour enhancers in cooked minced meat. *Food Chem.* **2013**, *141*, 77–83. [CrossRef]

20. Phat, C.; Moon, B.; Lee, C. Evaluation of umami taste in mushroom extracts by chemical analysis, sensory evaluation, and an electronic tongue system. *Food Chem.* **2016**, *192*, 1068–1077. [CrossRef]

21. Poojary, M.M.; Orlien, V.; Passamonti, P.; Olsen, K. Improved extraction methods for simultaneous recovery of umami compounds from six different mushrooms. *J. Food Compos. Anal.* **2017**, *63*, 171–183. [CrossRef]

22. Michael, H.W.; Bultosa, G.; Pant, L.M. Nutritional contents of three edible oyster mushrooms grown on two substrates at Haramaya, Ethiopia, and sensory properties of boiled mushroom and mushroom sauce. *Int. J. Food Sci. Technol.* **2011**, *46*, 732–738. [CrossRef]

23. Olawuyi, I.F.; Lee, W.Y. Quality and antioxidant properties of functional rice muffins enriched with shiitake mushroom and carrot pomace. *Int. J. Food Sci. Technol.* **2019**, *54*, 2321–2328. [CrossRef]

24. Gaglio, R.; Guarcello, R.; Venturella, G.; Palazzolo, E.; Francesca, N.; Moschetti, G.; Gargano, M.L. Microbiological, chemical and sensory aspects of bread supplemented with different percentages of the culinary mushroom *Pleurotus eryngii* in powder form. *Int. J. Food Sci. Technol.* **2019**, *54*, 1197–1205. [CrossRef]

25. Lu, X.; Brennan, M.A.; Serventi, L.; Mason, S.; Brennan, C.S. How the inclusion of mushroom powder can affect the physicochemical characteristics of pasta. *Int. J. Food Sci. Technol.* **2016**, *51*, 2433–2439. [CrossRef]

26. Novakovic, S.; Djekic, I.; Klaus, A.; Vunduk, J.; Djordjevic, V.; Tomovic, V.; Sojic, B.; Kocic-Tanackov, S.; Lorenzo, J.M.; Barba, F.; et al. The effect of *Cantharellus cibarius* addition on quality characteristics of frankfurter during refrigerated storage. *Foods* **2019**, *8*, 635. [CrossRef]

27. Fernandes, Â.; Barros, L.; Martins, A.; Herbert, P.; Ferreira, I.C.F.R. Nutritional characterisation of *Pleurotus ostreatus* (Jacq. ex Fr.) P. Kumm. produced using paper scraps as substrate. *Food Chem.* **2015**, *169*, 396–400. [CrossRef]

28. ISO. *International Standards Meat and Meat Products-Determination of Moisture Content: 1442*; International Organization for Standarization: Geneva, Switzerland, 1997.

29. ISO. *International Standards Meat and Meat Products-Determination of Ash Content: 936*; International Organization for Standarization: Geneva, Switzerland, 1998.

30. ISO. *International Standards Meat and Meat Products-Determination of Nitrogen Content: 937*; International Organization for Standarization: Geneva, Switzerland, 1978.

31. AOCS. *Official Procedure, Approved Procedure Am 5-04. Rapid Determination of Oil/Fat Utilizing High Temperature Solvent Extraction*; American Oil Chemists' Society: Urbana, IL, USA, 2005.

32. AOAC. *Official Methods of Analysis AOAC*, 15th ed.; AOAC: Washington, DC, USA, 1996.

33. Marti-Quijal, F.J.; Zamuz, S.; Tomašević, I.; Rocchetti, G.; Lucini, L.; Marszałek, K.; Lorenzo, J.M. A chemometric approach to evaluate the impact of pulses, *Chlorella* and *Spirulina* on proximate composition, amino acid, and physicochemical properties of turkey burgers. *J. Sci. Food Agric.* **2019**, *99*, 3672–3680. [CrossRef]

34. Vyncke, W. Evaluation of the Direct Thiobarbituric Acid Extraction Method for Determining Oxidative Rancidity in Mackerel (Scomber scombrus L.). *Fette Seifen Anstrichmittel* **1975**, *77*, 239–240. [CrossRef]

35. ISO. *Sensory Analysis e General Guidance for the Selection, Training and Monitoring of Assessors e Part 1: Selected Assessors, ISO 8586:201*; International Organization for Standardization: Geneva, Switzerland, 2012.

36. Ceron-Guevara, M.I.; Rangel-Vargas, E.; Lorenzo, J.M.; Bermúdez, R.; Pateiro, M.; Rodriguez, J.A.; Santos, E.M. Effect of the addition of edible mushroom flours (*Agaricus bisporus* and *Pleurotus ostreatus*) on physicochemical and sensory properties of cold-stored beef patties. *J. Food Process. Preserv.* **2020**, *44*, 14351. [CrossRef]

37. Carrasco-González, J.A.; Serna-Saldívar, S.O.; Gutiérrez-Uribe, J.A. Nutritional composition and nutraceutical properties of the *Pleurotus* fruiting bodies: Potential use as food ingredient. *J. Food Compos. Anal.* **2017**, *58*, 69–81. [CrossRef]

38. Elleuch, M.; Bedigian, D.; Roiseux, O.; Besbes, S.; Blecker, C.; Attia, H. Dietary fibre and fibre-rich by-products of food processing: Characterisation, technological functionality and commercial applications: A review. *Food Chem.* **2011**, *124*, 411–421. [CrossRef]

39. Figuerola, F.; Hurtado, M.L.; Estévez, A.M.; Chiffelle, I.; Asenjo, F. Fibre concentrates from apple pomace and citrus peel as potential fibre sources for food enrichment. *Food Chem.* **2005**, *91*, 395–401. [CrossRef]

40. Salcedo-Sandoval, L.; Cofrades, S.; Pérez, C.R.C.; Solas, M.T.; Jiménez-Colmenero, F. Healthier oils stabilized in konjac matrix as fat replacers in n-3 PUFA enriched frankfurters. *Meat Sci.* **2013**, *93*, 757–766. [CrossRef] [PubMed]

41. Reis, F.S.; Barros, L.; Martins, A.; Ferreira, I.C.F.R. Chemical composition and nutritional value of the most widely appreciated cultivated mushrooms: An inter-species comparative study. *Food Chem. Toxicol.* **2012**, *50*, 191–197. [CrossRef]

42. Cheung, P.C.K. Mini-review on edible mushrooms as source of dietary fiber: Preparation and health benefits. *Food Sci. Hum. Wellness* **2013**, *2*, 162–166. [CrossRef]

43. Corrêa, R.C.G.; Brugnari, T.; Bracht, A.; Peralta, R.M.; Ferreira, I.C.F.R. Biotechnological, nutritional and therapeutic uses of *Pleurotus* spp. (*Oyster mushroom*) related with its chemical composition: A review on the past decade findings. *Trends Food Sci. Technol.* **2016**, *50*, 103–117. [CrossRef]

44. Choi, Y.S.; Choi, J.H.; Han, D.J.; Kim, H.Y.; Lee, M.A.; Jeong, J.Y.; Kim, C.J. Effects of replacing pork back fat with vegetable oils and rice bran fiber on the quality of reduced-fat frankfurters. *Meat Sci.* **2010**, *84*, 557–563. [CrossRef]

45. Gasecka, M.; Siwulski, M.; Mleczek, M. Evaluation of bioactive compounds content and antioxidant properties of soil-growing and wood-growing edible mushrooms. *J. Food Process. Preserv.* **2018**, *42*, 1–10. [CrossRef]

46. González-Palma, I.; Escalona-Buendía, H.B.; Ponce-Alquicira, E.; Téllez-Téllez, M.; Gupta, V.K.; Díaz-Godínez, G.; Soriano-Santos, J. Evaluation of the antioxidant activity of aqueous and methanol extracts of *Pleurotus ostreatus* in different growth stages. *Front. Microbiol.* **2016**, *7*, 1099. [CrossRef]

47. Papastergiadis, A.; Mubiru, E.; Van Langenhove, H.; De Meulenaer, B. Malondialdehyde measurement in oxidized foods: Evaluation of the spectrophotometric thiobarbituric acid reactive substances (TBARS) test in various foods. *J. Agric. Food Chem.* **2012**, *60*, 9589–9594. [CrossRef]

48. Fernandes, R.P.P.; Trindade, M.A.; Lorenzo, J.M.; de Melo, M.P. Assesment of the stability of sheep sausages with the addition of different concentrations of *Origanum vulgare* extract during storage. *Meat Sci.* **2018**, *137*, 244–257. [CrossRef] [PubMed]

49. Wolfer, T.L.; Acevedo, N.C.; Prusa, K.J.; Sebranek, J.G.; Tarté, R. Replacement of pork fat in frankfurter-type sausages by soybean oil oleogels structured with rice bran wax. *Meat Sci.* **2018**, *145*, 352–362. [CrossRef] [PubMed]

50. Vilar, E.G.; Ouyang, H.; O'Sullivan, M.G.; Kerry, J.P.; Hamill, R.M.; O'Grady, M.N.; Mohammed, H.O.; Kilcawley, K.N. Effect of salt reduction and inclusion of 1% edible seaweeds on the chemical, sensory and volatile component profile of reformulated frankfurters. *Meat Sci.* **2020**, *161*, 108001. [CrossRef]

51. Jimenez-Colmenero, F.; Sánchez-Muniz, F.J.; Olmedilla-Alonso, B. Design and development of meat-based functional foods with walnut: Technological, nutritional and health impact. *Food Chem.* **2010**, *123*, 959–967. [CrossRef]

52. Mercadante, A.Z.; Capitani, C.D.; Decker, E.A.; Castro, I.A. Effect of natural pigments on the oxidative stability of sausages stored under refrigeration. *Meat Sci.* **2010**, *84*, 718–726. [CrossRef]

53. Stephan, A.; Ahlborn, J.; Zajul, M.; Zorn, H. Edible mushroom mycelia of *Pleurotus sapidus* as novel protein sources in a vegan boiled ssusage analog system: Functionality and sensory tests in comparison to commercial proteins and meat sausages. *Eur. Food Res. Technol.* **2018**, *244*, 913–924. [CrossRef]

54. Cofrades, S.; Serrano, A.; Ayo, J.; Solas, M.T.; Carballo, J.; Jiménez Colmenero, F. Restructured beef with different proportions of walnut as affected by meat particle size. *Eur. Food Res. Technol.* **2004**, *218*, 230–236. [CrossRef]

55. Miyaki, T.; Retiveau-Krogmann, A.; Byrnes, E.; Takehana, S. Umami increases consumer acceptability, and perception of sensory and emotional benefits without compromising health benefit perception. *J. Food Sci.* **2016**, *81*, S483–S493. [CrossRef]

 *foods*

Article

Application of Enoki Mushroom (*Flammulina Velutipes*) Stem Wastes as Functional Ingredients in Goat Meat Nuggets

Dipak Kumar Banerjee [1], **Arun K. Das** [1,*], **Rituparna Banerjee** [2], **Mirian Pateiro** [3], **Pramod Kumar Nanda** [1], **Yogesh P. Gadekar** [4], **Subhasish Biswas** [2], **David Julian McClements** [5] and **Jose M. Lorenzo** [3,*]

[1] Eastern Regional Station, ICAR-Indian Veterinary Research Institute, Kolkata 700 037, India; dipakkumarbanerjee23@gmail.com (D.K.B.); npk700@gmail.com (P.K.N.)

[2] Department of Livestock Products Technology, West Bengal University of Animal and Fishery Sciences, Kolkata 700 037, India; rituparnabnrj@gmail.com (R.B.); lptsubhasish@gmail.com (S.B.)

[3] Centro Tecnológico de la Carne de Galicia, Adva. Galicia n° 4, Parque Tecnológico de Galicia, San Cibrao das Viñas, 32900 Ourense, Spain; mirianpateiro@ceteca.net

[4] ICAR-Central Sheep and Wool Research Institute, Avikanagar, Jaipur 304 501, India; yogirajlpt@gmail.com

[5] Department of Food Science, University of Massachusetts, Armhrest, MA 01003, USA; mcclemen@umass.edu

* Correspondence: arun.das@icar.gov.in (A.K.D.); jmlorenzo@ceteca.net (J.M.L.)

Received: 4 March 2020; Accepted: 2 April 2020; Published: 4 April 2020

Abstract: The impact of different amounts (2%, 4% and 6%) of enoki (*Flammulina velutipes*) mushroom stem waste (MSW) powder on the physicochemical quality, color and textural, oxidative stability, sensory attributes and shelf-life of goat meat nuggets was evaluated. These mushroom by-products (MSW powder) contained a good source of protein (13.5%), ash (8.2%), total phenolics content (6.3 mg GAE/g), and dietary fiber (32.3%) and also exhibited the potential to be strong antioxidants, due to their good metal chelating ability (41.3%), reducing power (60.1%), and free radical scavenging activity (84.2%). Mushroom stem waste improved ($p < 0.05$) the emulsion stability, dietary fiber, ash and phenolics content of nuggets compared to control. Although no significant differences ($p > 0.05$) in expressible water and textural properties were observed among the formulations, but MSW powder improved the water holding capacity and slightly decreased the hardness. Further, the inclusion of MSW significantly ($p < 0.05$) improved the oxidative stability and shelf-life of treated nuggets by reducing lipid oxidation during the nine-day storage period. Again, the inclusion of MSW did not negatively affect the color and sensory attributes of treated meat nuggets. Overall, our results suggest that enoki mushroom stem waste (4%) can be used as a value-added functional ingredient to produce nutritionally improved and healthier meat products.

Keywords: enoki mushroom; physicochemical properties; antioxidant activity; dietary fiber; goat meat nuggets quality; sensory characteristics

1. Introduction

Mushrooms, due to their good nutritional attributes and richness in high-quality proteins, dietary fibers, vitamins, minerals, and phenolic compounds, are considered to be a healthy food product [1,2]. Enoki mushrooms (*Flammulina velutipes*), popularly known in different countries as "golden needle", "winter", "lily" or "velvet stem" mushrooms, are widely recognized for their good nutritional value and desirable taste attributes [3,4]. Several compounds including carbohydrates, protein, lipids, glycoproteins, phenols, and sesquiterpenes have been isolated from different parts of this mushroom [5]. Enoki mushrooms have also been known to exhibit good antioxidant, anti-inflammatory, immunomodulatory, anti-cancer and cholesterol-lowering activities [2,6]. The cultivated variety of this mushroom has a pure white bean sprout, a velvety stem, and a tiny snowy-white cap, whereas the wild variety may be orange to brown with a larger, shiny cap [7]. The stem base and other parts of the mushroom are removed during harvesting and these leftovers either go to landfills or are used as compost [8].

There is a growing interest in the application of plant-based waste materials as functional food ingredients in meat products, as they are a rich source of dietary fiber and several other bioactive compounds like vitamins, minerals, and polyphenols [9–12]. These dietary fibers, in combination with phenolic compounds, form antioxidant dietary fibers (ADFs) [13,14] which can be used as dietary supplements to improve gastrointestinal health, or as technical ingredients to inhibit lipid oxidation in foods, thereby extending their shelf-life [15,16]. As far as enoki mushrooms are concerned, its extract is reported to have strong antioxidant potential, with a high 2, 2-dipheny-1-picrylhydrazyl (DPPH) radical scavenging activity and metal chelating ability. Being a rich source of dietary fiber, the extract may reduce triacylglycerol, total cholesterol, and low-density lipoprotein levels in the blood due to a variety of mechanisms [4].

In summary, mushroom powder extracts have numerous nutritional attributes such as low calorie-density, healthy lipid profile, high fiber, protein and phenolic contents that make them suitable for incorporation as functional food ingredients in a variety of food products [17]. Previously, powdered mushroom extracts have been used to fortify a variety of foods, including biscuits, cookies, crackers, and cakes [18]. There have also been a limited number of studies on the incorporation of various kinds of mushroom extract into meat products. For instance, oyster mushrooms (*Pleurotus ostreatus*) have been investigated as a substitute for pork meat in Thai glutinous fermented sausage [19]. Reports on the impact of dried portobello mushroom (*Agaricus bisporus*) on the quality characteristics of a dry spicy sausage, sucuk [20], texture and structure of meat emulsions [1] and physicochemical and sensory properties of cold-stored beef patties [21], effects of oyster mushroom (*Pleurotus sajor-caju*) on the color, texture, cooking characteristics, and fiber content of chicken patties [22] are also available. Besides, enoki MSW has been used as a potential substitute for antibiotics in organic egg production by chickens [8] and to enhance the growth and health status of broiler chickens [3]. To the authors' knowledge, no study has previously been carried out on the utilization of enoki MSW as a functional food additive in processed meat products. The current study was designed to analyze the dietary fiber content and antioxidant properties of enoki MSW powder, and then to evaluate its potential as a functional ingredient in goat meat nuggets.

2. Materials and Methods

2.1. Materials and Reagents

Mushroom stem waste was collected from the harvesting and processing area of an enoki mushroom facility, cleaned properly for extraneous dirt, if any, and then dried in an oven (Static Oven, Instrumentation India, Kolkata) at 50 °C for 8 h. The dried MSW sample was then ground using a grinder (Kenstar, Mumbai, India) into powder (0.01 mm), which was then used as a functional ingredient in the goat meat product formulation. The proximate composition and antioxidant activity of this powder was also quantified. Goat meat (leg part) was procured from a supermarket and then kept in a freezer (−18 °C) until further processing. Chemical reagents such as methanol, sodium carbonate, 2-thiobarbituric acid (TBA), trichloroacetic acid (TCA), α-amylase, protease, amyloglucosidase, Folin-Ciocalteu (F-C) reagents, and gallic acid were procured from Sigma-Aldrich (Mumbai, India). Other chemicals and reagents were of analytical grade (SRL, Mumbai, India).

2.2. Chemical Composition and Extract Preparation of Mushroom Stem Waste

For chemical composition such as moisture, protein, fat, and ash content, duplicate MSW samples were analyzed based on the methods of the Association of Official Analytical Chemists [23]. The enzymatic-gravimetric process was used for dietary fiber estimation [24]. Briefly, different enzymes such as heat-stable α-amylase, protease and amyloglucosidase were used for sequential enzymatic digestion of MSW samples after its proper dispersion in phosphate buffer solution. Then, after filtration of the insoluble dietary fiber (IDF), warm distilled water was used to wash the residue. Ethanol (95%) was used for the precipitation of soluble dietary fiber (SDF) from the combined solution of filtrate and washings. The weight was noted after proper drying of residue in an oven (Static Oven, Instrumentation India, Kolkata) at 50 °C. These IDF and SDF residues later were also analyzed for protein and ash content. Both IDF and SDF fractions were combined for total dietary fiber (TDF) calculation.

For the preparation of MSW extract, water was used as a solvent [9]. Briefly, 20 g of MSW powder was accurately weighed into a conical flask. To this, 1000 mL of solvent was added and the whole content was held at room temperature (29 ± 1 °C) for 10 h, stirring frequently with a glass rod. The mixture was shaken at a constant rate (500 rpm) using a shaker, vortexed at high speed for 10 min, and finally centrifuged (REMI NEYA 8, Kolkata, India) at 5000× rpm for 10 min. The content of extract was then passed through a Whatman filter paper No. 1 (HiMedia®, Mumbai, India). The resulting extract was kept in a container and stored at 2 °C for further studies. The aqueous extracts obtained from repeated extractions were analyzed for total phenolic content (TPC), DPPH radical scavenging activity, ferrous ion chelating ability and reducing power assays. The efficacy of the extract was determined based on the dry weight of the mushroom powder.

2.3. Antioxidant Activity of Mushroom Stem Waste

2.3.1. Total Phenolics Content

The total phenolics content (TPC) of MSW was estimated using the Folin-Ciocalteu method. Briefly, 0.1 mL aqueous extract was properly mixed with 0.75 µL of F-C reagent and then a final volume of the above mix was increased ten-times using deionized water. Then, a sodium carbonate solution (750 µL) was added in each test tubes after 5 min and these tubs were incubated (in the dark) for 90 min at room temperature. The absorbance of test samples at 725 nm was taken using a spectrophotometer (Thermo Scientific, Wilmington, NC, USA) against a blank. Different concentrations of gallic acid were used for preparation of a standard curve and the TPC was calculated as gallic acid equivalents (GAE) in mg/g dry weight basis of MSW.

2.3.2. DPPH Radical Scavenging Activity

The method developed by Shimada et al. [25] was used for the measurement of DPPH radical scavenging activity. Briefly, 4 mL of methanol was added to a 1 mL extract from the MSW powder in a test tube and then 1 mL of 0.2 µM DPPH methanol solution was added and mixed. Samples were then incubated for 30 min and later the absorbance at 517 nm was measured using a spectrophotometer (Thermo Scientific, Wilmington, NC, USA). The scavenging activity was calculated by the following formula:

$$\text{Scavenging activity}(\%) = \left[1 - \left(\frac{absorbace\ of\ extract}{absorbance\ of\ control}\right) \times 100\% \right] \tag{1}$$

2.3.3. Ferrous Ion Chelating Ability

The ferrous ion chelating ability of the MSW extract was measured using the procedure outlined by Dinis et al. [26]. Briefly, 3.7 mL methanol and 0.1 mL of 2 μM FeCl$_2$, MSW extract (0.1 mL) were mixed properly and then held for 30 s before adding 0.1 mL of 5 mM ferrozine solution. The mixture was then kept for 10 min at room temperature for incubation purposes. Finally, the absorbance using a spectrophotometer (Thermo Scientific, Wilmington, NC, USA) was recorded at 562 nm. The formula for calculating the ferrous ion chelating ability is as follows:

$$\text{Chealting ability}(\%) = \left[1 - \left(\frac{absorbace\ of\ extract}{absorbance\ of\ control}\right) \times 100\% \right] \tag{2}$$

2.3.4. Ferric Reducing Antioxidant Power

The ferric reducing antioxidant power (FRAP) of the MSW extract was determined based on the method described by Madane et al. [9]. Briefly, 2.5 mL of extract taken in a 10 mL test tube was added with 2.5 mL of phosphate buffer (0.2 M, pH 6.6) and 2.5 mL of 1% (*w/v*) potassium ferricyanide. Then, 2.5 mL of 10% TCA was added after the mixture was kept for incubation at 50 °C for 20 min. After that, 2.5 mL of deionized water and 0.5 mL of ferric chloride (0.1% *w/v*) were mixed with 2.5 mL of the supernatant. Then, absorbance of samples was taken at 700 nm using a spectrophotometer (Thermo Scientific, Wilmington, NC, USA) and expressed as a percentage.

2.4. Preparation of Goat Meat Nuggets

Four formulations (control and treatments-T2, T4 and T6) of goat meat nuggets were prepared following the standard method described by Das et al. [27]. The first batch was considered as a control (meat without MSW powder), whereas in the case of T2, T4 and T6 formulations, MSW powder at 2.0%, 4.0% and 6.0% was included, respectively, replacing an equal percent of goat meat. Therefore, the total weight was 100 g with salt, condiments, spice mix, oil and wheat flour (Table 1).

Table 1. Formulation of goat meat nuggets with different levels of mushroom stem waste (MSW) powder.

Ingredients (%)	Treatment			
	Control	T2	T4	T6
Goat meat	71	69	67	65
Salt	1.5	1.5	1.5	1.5
Water (Ice)	10	10	10	10
Refined oil	8	8	8	8
Condiments *	4	4	4	4
Dry spice mix **	1.8	1.8	1.8	1.8
Wheat flour	3.5	3.5	3.5	3.5
Polyphosphate	0.03	0.03	0.03	0.03
Sodium nitrite (ppm)	150	150	150	150
MSW powder (%)	0.0	2.0	4.0	6.0

Treatments: Control = no additive; T2 = 2.0% MSW powder, T4 = 4.0% MSW powder and T6 = 6.0% MSW powder. * Condiments: fresh garlic and onion (4:1). ** Dry spice mix (18 g/kg nuggets): aniseed, black pepper, capsicum, caraway seed, cardamom, cinnamon, cloves, coriander powder, cumin seed, turmeric and dried ginger (Cookme, Kolkata, India).

Before processing, the frozen goat meat was thawed, cut into small cubes, and then minced using a meat mincer (Stadler, Mumbai, India). Meat emulsion was prepared separately for each group (control, T2, T4 and T6) by thoroughly mixing goat meat cubes with other ingredients (salt, sugar, phosphate, and nitrite) in a bowl chopper. During chopping, ice flakes were added to prevent excessive heating. Condiments, dry spice mix, and fine wheat flour were then added and chopped continuously until all the ingredients were uniformly mixed. About 500 g of emulsion from each formulation was placed in a mold, and cooked for 40 min with a steam-cooker (Stadler, Mumbai, India). The cooked meat blocks obtained were then sliced and cut into small pieces. Nugget samples of different formulations were then analyzed for various parameters (physicochemical, textural, colour) and also aerobically packed into low density polyethylene pouches and stored for up to nine days at 4 ± 1 °C for lipid oxidation study.

2.5. Analysis of Meat Products

2.5.1. pH, Emulsion Stability and Cooking Yield

The pH of the meat emulsion and nugget samples were measured after blending a 10 g sample with 50 mL of deionized water for a minute in a homogenizer (Omni International, Kennesaw GA, USA) and then using a digital pH meter. For emulsion stability, 25 g of emulsion was placed in a polypropylene bag and heated in a thermostatically controlled water bath for 20 min at 80 °C. The cooked mass was then cooled and weighed after draining out the exudate. For cooking yield, the weight of each meat block before and after cooking was recorded. The cooking yield was calculated and expressed as a percentage.

$$\text{Cooking yield (\%)} = \frac{\text{Weight of cooked meat block}}{\text{Weight of raw meat block}} \times 100 \tag{3}$$

2.5.2. Expressible Water

The percentage of expressible water which indicates the water holding capacity of cooked processed meat, was measured using the method of Madane et al. [9]. About 5 g of nugget sample was taken on two layers of Whatman No. 1 filter paper and placed the filter paper with sample in a 50 mL centrifuge tube for centrifugation (REMI NEYA 8, Kolkata, India) at 1500 rpm for 5 min. The sample was re-weighed after centrifugation and expressible water was calculated according to the following equation:

$$\text{Expressible water (\%)} = \frac{(\text{initial weight} - \text{final weight})}{\text{initial weight}} \times 100 \tag{4}$$

2.5.3. Total Phenolics Content in Meat Nuggets

Total phenolics content of meat nuggets was analyzed calorimetrically using F-C reagent. Briefly, the meat nugget sample (5 g) was homogenized at 3000 rpm for 2 min in a tube with 15 mL of distilled water. The mixture was shaken vigorously two to three times after adding 9 mL chloroform to separate the lipids. The F-C reagent (500 μL) was added to 1 mL aliquot of the diluted sample (1:4, *v/v*) followed by the addition of 1 mL of the sodium carbonate solution (5%). After proper mixing, samples were incubated for 1 h at room temperature and then vortexed. Then absorbance was taken at 700 nm. The results were expressed as mg gallic acid equivalents (GAE)/g of the dry weight of the meat nuggets.

2.5.4. Texture Profile Analysis

The textural properties such as hardness, springiness, gumminess, cohesiveness and chewiness of the control and MSW powder treated goat meat nuggets were measured with the help of a texture analyzer (Stable Micro System Model TA.XT 2i/25, Surrey, UK). The measurement was carried out using central cores of five pieces of each meat sample (2 cm × 2 cm × 2 cm). Each sample was compressed twice (80% of the original height and 2 mm/s crosshead speed) with the help of a compression probe (P 75).

2.5.5. Instrumental Color Attributes

Various color attributes of the meat nuggets were measured using a Hunter Color Lab instrument in which the Hunter L*, a*, and b* values were determined. L* denoted pure white (100)/pure black (0), a* +redness/-greenness and b* +yellowness/-blueness. A light trap/black glass and white tile was used for calibration of instrument. The color attributes of the nugget surfaces were analyzed at three different points.

2.5.6. Sensory Evaluation

The sensory attributes, including appearance, flavor, texture, juiciness, and overall acceptability, of meat nuggets were evaluated using an eight-point descriptive scale, where 8 = excellent and 1 = extremely poor [28]. The purpose of the experiments was clearly explained to the panelists without disclosing the much about samples' identity. The panelists evaluated the sample based on a sensory preformed provided after coding with number and warming in a microwave oven for 1 min. Water was provided to the panelists to rinse their mouths during evaluation.

2.6. Lipid Oxidation in Meat Products

2.6.1. Peroxide Values

The peroxide values (PV) of the nugget samples were determined following the procedure described by Koniecko [29] with slight modifications. Briefly, for filtrate preparation, anhydrous sodium sulfate and chloroform was used to blend the meat sample (5 g). 2 mL of saturated potassium iodide solution were added to a mixture of 25 mL filtrate and 30 mL glacial acetic acid. After proper mixing for about 2 min, 100 mL of distilled water and 2 mL of fresh 1% starch solution were added and the mixture was titrated immediately with 0.1 N sodium thiosulphate until the end point was reached (the non-aqueous layer turned colorless). The PV of the meat sample was calculated and expressed in meqO$_2$/kg of the sample.

2.6.2. Thiobarbituric Acid Reacting Substances

The measurement of thiobarbituric acid reacting substances (TBARS) was carried out to measure the lipid oxidation in the meat nugget during storage [30]. Briefly, the TCA extract was prepared after triturating samples with 20% pre-cooled TCA (25 mL). To 3 mL TCA extract (filtrate), 3 mL of TBA reagent (5 mM) was added and then cooled in running tap water after boiling in a thermostatically controlled water bath at 70 °C for 35 min. Similarly, a blank was prepared by adding and properly mixing 3 mL of 10% TCA and 3 mL of the 5mM TBA reagent. Then, absorbance was recorded at a 532 nm. The TBARS value was expressed as mg malonaldehyde (MDA) per kg of meat sample.

2.7. Statistical Analysis

Measurements of all the parameters were performed in duplicate and this study was conducted thrice. The data collected from this study were analyzed by a SPSS software (version 20.0, Armonk, NY, USA). The normal distribution and variance homogeneity were previously assessed (Shapiro-Wilk). In the case of the lipid oxidation study, the data were analyzed using two-way ANOVA with treatments (control, T2, T4 and T6) and storage time (0, 3, 6, 9 days) as the main effects (4 × 4 factorial design). To find out the impact of MSW on various parameters, Duncan's multiple range tests were applied. The values are presented here as the mean and standard error, and significance differences were identified at the 95% confidence level (P < 0.05).

3. Results and Discussion

3.1. Proximate Composition and Dietary Fiber Content of Mushroom Stem Waste

The proximate composition and dietary fiber content of enoki MSW powder is presented in Table 2. The MSW powder had high moisture (12.9%), protein (13.5%) and ash (8.2%), but a relatively lower fat content (1.5%). Available reports indicate that dried mushrooms contain relatively high moisture levels, depending on the species and other factors [31], more than 25% protein, less than 3% crude fat, and around 50% total carbohydrate [32]. Further, the ash content in mushrooms typically ranges between 5%–12% of dry matter. The composition of enoki MSW analyzed by other researchers [3,8,33] varied between 12.75% and 18.42% for crude protein, fat from 1.5% to 2.94% and ash between 6.33% and 11.6%, which is well within the level found in our study (Table 2). In this regard, variation in proximate composition could be due to differences in harvesting methods and stages of maturity of mushroom, soil types and environmental factors [34].

In this study, the SDF and IDF contents of the enoki MSW were found to be 17.3% and 15.1%, respectively (Table 2). Many researchers have reported the SDF and IDF content of mushrooms ranging between 22.4–31.2% and 4.2–9.2% of dry weight, respectively [35]. The TDF (32.3%) content determined in our study is fairly similar to that reported for raw enoki mushrooms (29.3%) [36]. In a comparison of different kinds of mushrooms, Yang, Lin, & Mau [37] found a higher dietary fiber content in enoki mushrooms than in shiitake or oyster mushrooms. The ability of enoki mushrooms to lower cholesterol and blood pressure levels may be partly attributed to their relatively high dietary fiber content [4]. Moreover, these dietary fibers may help in the formulation of low-calorie, low-fat and high-fiber meat products, due to their ability to form gel networks that hold water and modulate texture.

Table 2. Proximate composition (mean values ± SE) and antioxidant activity of enoki mushroom stem waste powder/extract.

Proximate Composition (g/100 g Dry Matter)	
Moisture	12.9 ± 0.3
Protein	13.5 ± 0.7
Fat	1.47 ± 0.04
Ash	8.24 ± 0.05
Total dietary fiber	32.3 ± 0.9
Soluble dietary fiber	17.3 ± 2.1
Insoluble dietary fiber	15.1 ± 2.7
Antioxidant Capacity of MSW Extract (1 mg/mL)	
Total phenolics (mg GAE/g)	6.3 ± 2.5
DPPH scavenging (%)	84.2 ± 3.0
FRAP (%)	60.1 ± 1.2
Chelating ability of ferrous ion (%)	41.3 ± 0.5

DPPH: 2, 2-diphenyl-1- picrylhydrazyl; FRAP: ferric reducing antioxidant power; GAE: gallic acid equivalents; $n = 6$.

3.2. Antioxidant Activity of Mushroom Stem Waste Extract

The total phenolics content of the enoki mushroom stem extract was 6.26 mg GAE/g dry weight, which was determined using gallic acid as a standard (Table 2). The antioxidant potential of this extract was assessed using a number of assays: the DPPH assay was used to measure the free radical scavenging ability [38]; the FRAP assay was used to measure the reducing power (Fe^{3+} to Fe^{2+}); and, the iron-binding assay was used to measure the ability to chelate transition metal irons [39]. The enoki mushroom stem extract was found to have strong antioxidative potential: 84.2% DPPH scavenging; 60.1% reducing power; and 41.3% of ferrous ion chelating ability (Table 2).

In fact, phenolic compounds that are found naturally in mushrooms have antioxidant activity due to their hydrogen-donating and singlet oxygen-quenching properties [40,41]. The antioxidant properties of the enoki mushroom extract can be attributed to a number of antioxidant constituents, including p-coumaric acid, ellagic acid [42], gallic acid, pyrogallol, chlorogenic acid, caffeic acid, ferulic acid, and quercetin [40]. The phenolics content of mushrooms has been shown to be positively correlated with the results of the DPPH assay and other antioxidant assays [4]. Previous studies have shown that enoki mushrooms have higher phenolics contents, ferric reducing powers, and ferrous chelating activities than other mushrooms [43]. Taken together, these studies suggest that enoki mushroom extracts are a good source of natural antioxidants.

3.3. Physicochemical Properties and Proximate Composition of Fortified Goat Meat Nuggets

The pH of the meat emulsion without MSW powder (control) was the lowest among all the treatments (Table 3). The addition of the mushroom powder significantly ($p < 0.05$) increased the pH. The meat emulsion containing 6.0% MSW had the highest pH value (6.44). Our results are in agreement with the findings of Bao, Ushio, & Ohshima [44], who earlier reported that the addition of enoki mushroom extracts to beef and fish slightly increased the pH of their products, although the increase was statistically non-significant. The increase in pH of products could be due to the abundance of basic amino acids in comparison to acidic amino acids with addition of enoki mushroom powder [45], as well as the natural buffering capacity of the mushroom proteins [46].

Table 3. Effect of enoki mushroom stem waste (MSW) on the physicochemical, textural and color attributes of goat meat nuggets ($n = 6$).

Parameters	Treatments			
	Control	T2	T4	T6
Emulsion pH	6.33 ± 0.02 [c]	6.37 ± 0.02 [bc]	6.39±0.02 [a]	6.40±0.02 [a]
Emulsion Stability (%)	94.32 ± 0.20 [b]	95.86 ± 0.19 [a]	96.25±0.22 [a]	96.67±0.22 [a]
pH	6.38 ± 0.01 [c]	6.40 ± 0.01 [bc]	6.42±0.10 [a]	6.44±0.1 [a]
Cooking loss (%)	5.08 ± 0.12 [c]	4.12 ± 0.08 [b]	3.83 ± 0.16 [ab]	3.12 ± 0.18 [a]
Total phenolics content (mg GAE/g)	0.142 ± 0.42 [d]	0.44 ± 0.39 [c]	0.62 ± 0.45 [b]	0.96 ± 0.38 [a]
Expressible water (%)	26.3 ± 1.4	24.2 ± 2.3	22.4 ± 2.4	21.84 ± 2.02
Proximate Composition (g/100 g)				
Moisture	65.29 ± 0.54	65.36 ± 0.82	66.74 ± 0.56	67.23 ± 0.56
Protein	15.35 ± 0.24	15.17 ± 0.29	15.22 ± 0.26	14.68 ± 0.20
Fat	12.26 ± 0.29	12.15 ± 0.24	12.13 ± 0.32	12.08 ± 0.12
Ash	2.67 ± 0.05 [d]	3.54 ± 0.03 [c]	4.26 ± 0.05 [b]	4.68 ± 0.03 [a]
Total dietary fiber	0.82 ± 0.06 [d]	1.28 ± 0.04 [c]	1.42 ± 0.06 [b]	1.72 ± 0.04 [a]
Textural Parameters				
Hardness (N/cm^2)	42.42 ± 1.86	38.40 ± 1.92	36.33 ± 2.08	34.33 ± 2.12
Springiness (cm)	0.86 ± 0.01	0.84 ± 0.02	0.83 ± 0.02	0.83 ± 0.02
Cohesiveness	0.48 ± 0.02	0.47 ± 0.01	0.45 ± 0.01	0.44 ± 0.01
Gumminess (N/cm^2)	14.79 ± 1.04	13.42 ± 1.22	12.83 ± 1.34	12.08 ± 1.57
Chewiness (N/cm)	14.05 ± 0.82 [a]	13.14 ± 0.78 [ab]	11.75 ± 0.91 [b]	9.46 ± 0.84 [b]
Color Parameters				
L* value	47.42 ± 0.22 [c]	48.68 ± 0.19 [bc]	49.42 ± 0.24 [ab]	52.02 ± 0.20 [a]
a* value	7.62 ± 0.20 [a]	7.48 ± 0.18 [a]	7.22 ± 0.16 [ab]	6.32 ± 0.14 [b]
b* value	13.24 ± 0.32	13.12 ± 0.28	13.18 ± 0.22	13.20 ± 0.24

Treatments: Control = no additive; T2 = 2.0% MSW powder, T4 = 4.0% MSW powder and T6 = 6.0% MSW powder.
[a–c] Mean values in the same row bearing different superscript differ significantly ($p < 0.05$).

There was a significant ($p < 0.05$) increase in the emulsion stability of the treated meat nuggets, whereas cooking loss (%) reduced significantly with an increase in level of MSW powder incorporation, compared to control. On the other hand, the expressible water (%) decreased with increased powder level, indicating an improvement in water holding capacity, although this change was not statistically significant ($p > 0.05$). The improved emulsion stability and reduced cooking loss were probably because of the higher TDF (32.3%) content of enoki MSW which enhanced the oil absorption and water retention properties of the meat emulsion [47]. The improvement in water binding and fat retention in meat products, upon the addition of dietary fiber from several sources, have been reported by various researchers [16,47].

The TPC of meat nuggets increased significantly ($p < 0.05$) with the increasing level of MSW powder, rising from 0.14 to 0.96 mg GAE/g dry weight of product as the MSW content increased up to 6.0%. This is in agreement with the findings of different researchers who reported significantly increased TPC of meat nuggets upon addition of guava powder [16] and dragon fruit peel powder [9]. The increased phenolics content in treated meat nuggets could be due to addition of powdered enoki mushroom stem extract, which is reported to possess several bioactive phenolic and polyphenolic compounds [40,42,43].

Incorporation of MSW powder did not have a significant effect on the moisture, protein, and fat contents of the meat nuggets, but it led to a significant ($p < 0.05$) increase in the ash and dietary fiber contents (Table 3). These effects can be attributed to the presence of relatively high levels of minerals and dietary fibers in the mushrooms. It has been reported that this kind of mushroom is not only rich in potassium and phosphorus [43,48], but also contains several other minerals in minor amounts such as sulfur, sodium, copper, iron, and zinc [31]. In addition, the powdered enoki mushroom extract contained higher levels of TDF (32%), which would contribute to the fiber content of the final meat product.

3.4. Textural and Color Attributes of Meat Nuggets

The effects of incorporating powdered enoki MSW on the textural and color attributes of the meat nuggets is presented in Table 3. Although various textural parameters like hardness, springiness, cohesiveness, and gumminess of the nuggets decreased slightly, the reduction was not statistically significant. However, the chewiness of treated nuggets decreased significantly ($p < 0.05$) with an increased level of mushroom powder addition. Our results are in tandem with the findings of Choe et al. [49] who added enoki mushroom extracts to emulsion-type sausages and obtained similar results. Other researchers have shown that replacing chicken meat with 25% or 50% oyster mushroom (*P. sajor-caju*) reduced the hardness and improved the textural parameters of chicken patties [22]. The addition of portobello mushroom powder has been shown to increase the hardness and cohesiveness of model meat emulsions up to 3%, but then decrease their textural attributes at higher addition levels [1]. The textural properties of cooked meat products containing plant materials are related to the gelation of the myofibrillar proteins from the meat [50], as well as the biopolymer networks formed by the dietary fibers from the plants [47]. The addition of dietary fiber may have influenced the gelation of the meat proteins, thereby decreasing the gel strength and leading to a softer texture [14,15,47].

The color of any fresh or processed food product plays an important role in influencing the decision of consumers [51]. It is, therefore, imperative that any functional ingredient added to improve the nutritional properties of a meat product does not cause undesirable changes in its appearance. Increasing the amount of powdered enoki mushroom extract in nugget formulations increased the lightness (L*) and reduced redness (a*), but did not change their yellowness (b*) values (Table 3). No statistical differences ($p > 0.05$) in the lightness and redness of the meat nuggets containing 2.0% and 4.0% mushroom extract were observed compared to the control group. However, there was a significant increase in lightness and reduction in redness ($p < 0.05$) in nuggets samples prepared with 6.0% MSW. This might be a result of the dilution of the meat protein due to the addition of mushroom powder as a percentage of the meat and the white color of mushroom powder. Moreover, there may have occurred an increase in the degree of light scattering by the particles in the mushroom powder, which caused the meat product to become lighter. Similar results were reported by Choe et al. [49] in enoki mushroom powder added to emulsion-type sausages. In another study, it was reported that adding portobello mushroom extracts to a meat emulsion led to a decrease in L* value and an increase in a* value [1], which suggests that the effects may be system dependent.

3.5. Sensory Characteristics of Meat Nuggets

The sensory parameters of goat meat nuggets containing different levels of powdered enoki mushroom extract are presented in Table 4. There was no significant difference ($p > 0.05$) in the individual sensory attributes of the meat nuggets regardless of the level of mushroom waste used. However, the appearance, flavor, and overall acceptability of nuggets decreased slightly with the addition of 6.0% MSW powder. Unlike other mushroom species, enoki mushroom is reported to have a very mild and delicate taste [52]. This is desirable and may be a beneficial attribute for many food applications, as it does not strongly alter or mask the expected sensory attributes of the original product. Few reports available in this regard suggest that enoki mushrooms contain high levels of free amino acids, which are associated with umami or monosodium glutamate-like, sweet, and bitter tastes that are often perceived favorably by the consumers [37]. The white color of the mushrooms may also be beneficial because it does not alter the overall hue of the final meat product, but may slightly decrease its lightness.

In this regard, researchers have shown that supplementation of pork patties with the ground, white jelly mushrooms at 10%, 20%, and 30% by weight did not affect the liking of appearance, color, flavor, or texture [53], but the acceptability was better at a 10% level. In another study, Myrdal Miller et al. [54] reported that the addition of ground white button mushrooms to ground meat did not have a major impact on their perceived quality attributes. Taken together, these results suggest that mushrooms can be used as a healthy substitute for meat products without adversely impacting their desirable appearance or flavor.

Table 4. Effects of enoki mushroom stem waste (MSW) powder on the sensory characteristics of goat meat nuggets ($n = 30$).

Parameters	Treatments			
	Control	T2	T4	T6
Appearance	7.03 ± 0.12	7.07 ± 0.10	7.00 ± 0.14	6.94 ± 0.12
Texture	6.88 ± 0.13	6.92 ± 0.16	7.01 ± 0.10	6.96 ± 0.14
Flavor	6.84 ± 0.10	6.90 ± 0.18	6.86 ± 0.16	6.80 ± 0.14
Juiciness	6.74 ± 0.11	6.89 ± 0.13	6.98 ± 0.12	6.95 ± 0.12
Overall acceptability	6.87 ± 0.14	6.90 ± 0.16	6.98 ± 0.18	6.84 ± 0.20

Treatments: Control = no additive; T2 = 2.0% MSW powder, T4 = 4.0% MSW powder and T6 = 6.0% MSW powder.

3.6. Lipid Oxidation of Meat Nuggets During Storage

The impact of the powdered MSW on the oxidative stability of the meat nuggets was determined by measuring the primary (PV) and secondary (TBARS) lipid oxidation products over time and is presented in Figures 1 and 2. The treated meat nuggets (T2, T4, and T6) had significantly ($p < 0.05$) lower peroxide and TBARS values than the control group, although there was no significant difference ($p > 0.05$) between the levels of powder addition.

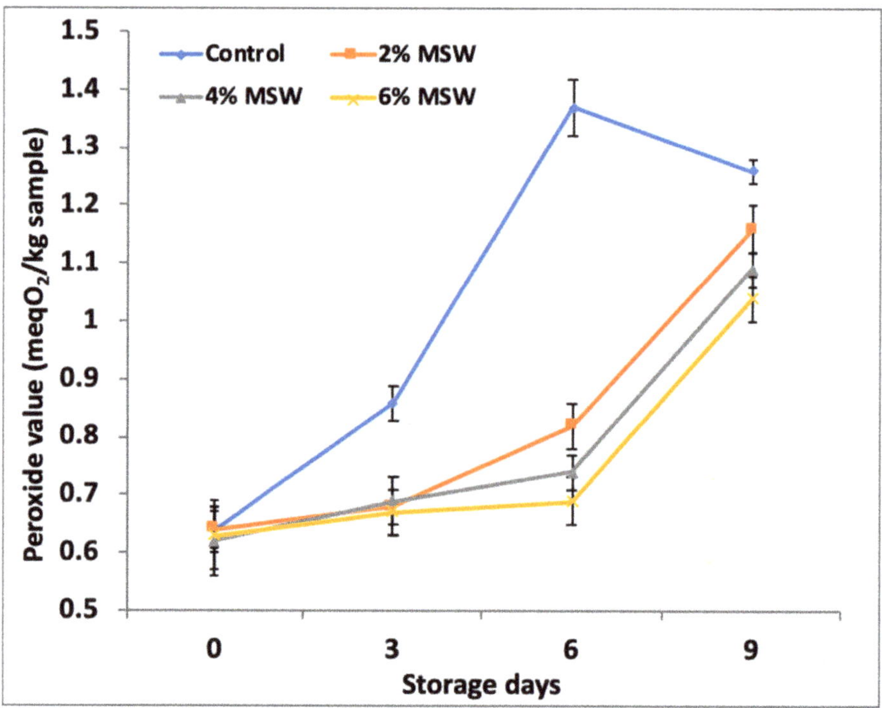

Figure 1. Effect of mushroom stem waste (MSW) powder on peroxide values of goat meat nuggets during storage.

Hydroperoxides are primary products of lipid oxidation, hence PV are important to know the extent of initial lipid oxidation in meat samples. The results depicted in Figure 1 indicated that the initial PV (0.64 meqO$_2$/kg) of control nuggets increased to 1.21 meqO$_2$/kg after nine days of storage, which was significantly higher ($p < 0.05$) compared to treated nuggets. Similarly, results suggested that the control samples underwent noticeable lipid oxidation during the first six days of refrigerated storage and reached maximum PV at the end of the primary auto-oxidation. After six days of storage, the hydroperoxides formed might have gone through the decomposition to form secondary lipid oxidation products [55]. Although oxidation in the control was more intense compared to the treated samples, a decline was observed on day six. This indicates that, after the induction period, the decomposition rate of the hydroperoxides was faster than the production rate [56].

The control nugget had an initial TBARS value of 0.32 and it reached 0.85 mg MDA/kg on the ninth day of storage study, whereas TBARS value in treated nuggets with 2–6% MW increased from 0.32–0.58 mg MDA/kg (Figure 2). There was an increase in TBARS values during storage irrespective of treatment but at a slower rate in treated nuggets compared to the control, indicating the effectiveness of MSW in inhibiting lipid oxidation during the storage. MSW besides supplementing the dietary fiber to goat meat nuggets was found to retard lipid peroxidation in the product during refrigerated storage. Although the secondary reaction products showed an upward trend in all the samples as storage days progressed, a considerably slower rate was observed in the samples containing the mushroom extracts. These results suggest that the enoki mushroom extracts were effective antioxidants that were able to retard lipid peroxidation in treated goat meat nuggets during refrigerated storage study for up to nine days.

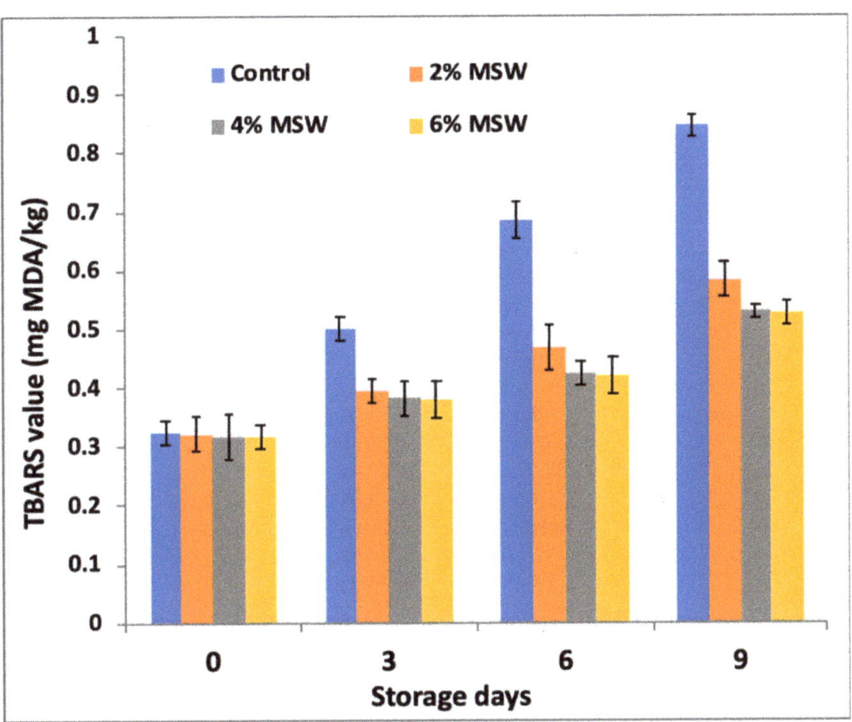

Figure 2. Effect of mushroom stem waste (MSW) powder on TBARS values of goat meat nuggets during storage.

In general, mushrooms have been shown to contain a variety of different antioxidant substances that make them effective at inhibiting lipid oxidation [4,7,31,57]. In an earlier section, we showed that the enoki mushroom extracts had strong reducing power, high scavenging activity, and good iron-binding ability (Table 2). These antioxidant effects may be attributed to the relatively high phenolic, dietary fiber, ergothioneine, vitamin C, and nucleotides content [7]. Antioxidant effects have also been reported when winter mushroom extract is added to beef and fish products [44] and emulsion-type sausages [49]. Enoki mushroom extracts have also been shown to prevent discoloration and lipid oxidation in fish and melanosis in crustaceans during postmortem storage [44,58].

4. Conclusions

Our study has shown that enoki mushroom (*F. velutipes*) stem waste is a good source of bioactive ingredients, such as dietary fibers and phenolics. The extract of MSW powder exhibited good antioxidant potential, which was attributed to its strong free radical scavenging activity, ferric reducing power, and metal chelating ability. The incorporation of this extract into a model meat product (goat meat nuggets) increased its dietary fiber and ash content, which may have nutritional benefits. Moreover, the mushroom extract improved the cooking yield and did not adversely affect the appearance or texture of the final product. Finally, the mushroom extract significantly improved the shelf-life of the meat products, which was mainly attributed to its ability to inhibit lipid oxidation during storage. Therefore, enoki mushroom stem waste may be a value-added functional ingredient that can be used at 4% level to improve the nutritional profile, physicochemical properties, and shelf-life of meat products.

Author Contributions: Conceptualization, A.K.D., and D.K.B.; Investigation, analysis and interpretation of the data, writing—original draft preparation, D.K.B., R.B., and Y.P.G.; writing—review and editing, P.K.N., A.K.D., S.B., M.P., D.J.M., and J.M.L. All authors have read and agreed to the published version of the manuscript.

Funding: This research work received no external funding.

Acknowledgments: The authors are thankful to the Director, ICAR-Indian Veterinary Research Institute (IVRI), Bareilly, India and the Station In-charge, ICAR-IVRI, Kolkata for providing necessary facilities to carry out this work.

Conflicts of Interest: The authors declare no conflict of interest.

References

1. Kurt, A.; Gençcelep, H. Enrichment of meat emulsion with mushroom (*Agaricus bisporus*) powder: Impact on rheological and structural characteristics. *J. Food Eng.* **2018**, *237*, 128–136. [CrossRef]

2. Valverde, M.E.; Hernández-Pérez, T.; Paredes-López, O. Edible mushrooms: Improving human health and promoting quality life. *Int. J. Microbiol.* **2015**, *2015*. [CrossRef]

3. Mahfuz, S.; He, T.; Liu, S.; Wu, D.; Long, S.; Piao, X. Dietary inclusion of mushroom (*Flammulina velutipes*) stem waste on growth performance, antibody response, immune status, and serum cholesterol in broiler chickens. *Animals* **2019**, *9*, 692. [CrossRef]

4. Yeh, M.Y.; Ko, W.C.; Lin, L.Y. Hypolipidemic and antioxidant activity of enoki mushrooms (*Flammulina velutipes*). *Biomed Res. Int.* **2014**, *2014*. [CrossRef]

5. Yang, W.; Fang, Y.; Liang, J.; Hu, Q. Optimization of ultrasonic extraction of *Flammulina velutipes* polysaccharides and evaluation of its acetylcholinesterase inhibitory activity. *Food Res. Int.* **2011**, *44*, 1269–1275. [CrossRef]

6. Wu, D.M.; Duan, W.Q.; Liu, Y.; Cen, Y. Anti-inflammatory effect of the polysaccharides of golden needle mushroom in burned rats. *Int. J. Biol. Macromol.* **2010**, *46*, 100–103. [CrossRef]

7. Tang, C.; Hoo, P.C.X.; Tan, L.T.H.; Pusparajah, P.; Khan, T.M.; Lee, L.H.; Goh, B.H.; Chan, K.G. Golden needle mushroom: A culinary medicine with evidenced-based biological activities and health promoting properties. *Front. Pharmacol.* **2016**, *7*, 474. [CrossRef]

8. Mahfuz, S.; Song, H.; Liu, Z.; Liu, X.; Diao, Z.; Ren, G.; Guo, Z.; Cui, Y. Effect of golden needle mushroom (*Flammulina velutipes*) stem waste on laying performance, calcium utilization, immune response and serum immunity at early phase of production. *Asian Australas. J. Anim. Sci.* **2018**, *31*, 705–711. [CrossRef]

9. Madane, P.; Das, A.K.; Pateiro, M.; Nanda, P.K.; Bandyopadhyay, S.; Jagtap, P.; Barba, F.J.; Shewalkar, A.; Maity, B.; Lorenzo, J.M. Drumstick (*Moringa oleifera*) flower as an antioxidant dietary fibre in chicken meat nuggets. *Foods* **2019**, *8*, 307. [CrossRef]

10. Lorenzo, J.M.; Pateiro, M.; Domínguez, R.; Barba, F.J.; Putnik, P.; Kovačević, D.B.; Shpigelman, A.; Granato, D.; Franco, D. Berries extracts as natural antioxidants in meat products: A review. *Food Res. Int.* **2018**, *106*, 1095–1104. [CrossRef]

11. De Carvalho, F.A.L.; Lorenzo, J.M.; Pateiro, M.; Bermúdez, R.; Purriños, L.; Trindade, M.A. Effect of guarana (*Paullinia cupana*) seed and pitanga (*Eugenia uniflora* L.) leaf extracts on lamb burgers with fat replacement by chia oil emulsion during shelf life storage at 2 °C. *Food Res. Int.* **2019**, *125*, 1–10. [CrossRef]

12. Pateiro, M.; Vargas, F.C.; Chincha, A.A.I.A.; Sant'Ana, A.S.; Strozzi, I.; Rocchetti, G.; Barba, F.J.; Domínguez, R.; Lucini, L.; do Amaral Sobral, P.J.; et al. Guarana seed extracts as a useful strategy to extend the shelf life of pork patties: UHPLC-ESI/QTOF phenolic profile and impact on microbial inactivation, lipid and protein oxidation and antioxidant capacity. *Food Res. Int.* **2018**, *114*, 55–63. [CrossRef]

13. Saura-Calixto, F. Antioxidant dietary fiber product: A new concept and a potential food ingredient. *J. Agric. Food Chem.* **1998**, *46*, 4303–4306. [CrossRef]

14. Das, A.K.; Nanda, P.K.; Madane, P.; Biswas, S.; Das, A.; Zhang, W.; Lorenzo, J.M. A comprehensive review on antioxidant dietary fibre enriched meat-based functional foods. *Trends Food Sci. Technol.* **2020**, *99*, 323–336. [CrossRef]

15. Madane, P.; Das, A.K.; Nanda, P.K.; Bandyopadhyay, S.; Jagtap, P.; Shewalkar, A.; Maity, B. Dragon fruit (*Hylocereus undatus*) peel as antioxidant dietary fibre on quality and lipid oxidation of chicken nuggets. *J. Food Sci. Technol.* **2020**, *57*, 1449–1461. [CrossRef]

16. Verma, A.K.; Rajkumar, V.; Banerjee, R.; Biswas, S.; Das, A.K. Guava (*Psidium guajava* L.) powder as an antioxidant dietary fibre in sheep meat nuggets. *Asian Australas. J. Anim. Sci.* **2013**, *26*, 886. [CrossRef]

17. Abugri, D.A.; McElhenney, W.H. Fatty acid profiling in selected cultivated edible and wild medicinal mushrooms in southern United States. *J. Exp. Food Chem.* **2016**, *2*, 1–7. [CrossRef]

18. Reis, F.S.; Martins, A.; Vasconcelos, M.H.; Morales, P.; Ferreira, I.C.F.R. Functional foods based on extracts or compounds derived from mushrooms. *Trends Food Sci. Technol.* **2017**, *66*, 48–62. [CrossRef]

19. Chockchaisawasdee, S.; Namjaidee, S.; Pochana, S.; Stathopoulos, C.E. Development of fermented oyster-mushroom sausage. *Asian J. Food Agro Ind.* **2010**, *3*, 35–43.

20. Gençcelep, H. The effect of using dried mushroom (*Agaricus bisporus*) on lipid oxidation and color properties of sucuk. *J. Food Biochem.* **2012**, *36*, 587–594. [CrossRef]

21. Cerón-Guevara, M.I.; Rangel-Vargas, E.; Lorenzo, J.M.; Bermúdez, R.; Pateiro, M.; Rodriguez, J.A.; Sanchez-Ortega, I.; Santos, E.M. Effect of the addition of edible mushroom flours (*Agaricus bisporus* and *Pleurotus ostreatus*) on physicochemical and sensory properties of cold-stored beef patties. *J. Food Process. Preserv.* **2019**. [CrossRef]

22. Wan Rosli, W.I.; Solihah, M.A.; Aishah, M.; Nik Fakurudin, N.A.; Mohsin, S.S.J. Colour, textural properties, cooking characteristics and fibre content of chicken patty added with oyster mushroom (*Pleurotus sajor-caju*). *Int. Food Res. J.* **2011**, *18*, 621–627.

23. AOAC. *Official Methods of Analysis*, 16th ed.; Association of Official Analytical Chemists: Washington, DC, USA, 1995.

24. Prosky, L.; Asp, N.-G.; Schweizer, T.F.; DeVries, J.W.; Furda, I. Determination of insoluble, soluble, and total dietary fiber in foods and food products: Interlaboratory study. *J. Assoc. Off. Anal. Chem.* **1988**, *71*, 1017–1023. [CrossRef]

25. Shimada, K.; Fujikawa, K.; Yahara, K.; Nakamura, T. Antioxidative properties of xanthan on the autoxidation of soybean oil in cyclodextrin emulsion. *J. Agric. Food Chem.* **1992**, *40*, 945–948. [CrossRef]

26. Dinis, T.C.P.; Madeira, V.M.C.; Almeida, L.M. Action of phenolic derivatives (acetaminophen, salicylate, and 5-aminosalicylate) as inhibitors of membrane lipid peroxidation and as peroxyl radical scavengers. *Arch. Biochem. Biophys.* **1994**, *315*, 161–169. [CrossRef]

27. Das, A.K.; Rajkumar, V.; Verma, A.K. Bael pulp residue as a new source of antioxidant dietary fiber in goat meat nuggets. *J. Food Process. Preserv.* **2015**, *39*, 1626–1635. [CrossRef]

28. Chauhan, P.; Das, A.K.; Das, A.; Bhattacharya, D.; Nanda, P.K. Effect of black cumin and arjuna fruit extract on lipid oxidation in pork nuggets during refrigerated storage. *J. Meat Sci.* **2018**, *13*, 73–80. [CrossRef]

29. Koniecko, E.S. *Handbook for Meat Chemists*; Avery Publishing Group Inc.: Wayne, NJ, USA, 1979; ISBN 089529060X.

30. Witte, V.C.; Krause, G.F.; Bailey, M.F. A new extraction method for determining 2-thiobarbituric acid values of pork and beef during storage. *J. Food Sci.* **1970**, *35*, 582–585. [CrossRef]

31. Guillamón, E.; García-Lafuente, A.; Lozano, M.; Dóarrigo, M.; Rostagno, M.A.; Villares, A.; Martínez, J.A. Edible mushrooms: Role in the prevention of cardiovascular diseases. *Fitoterapia* **2010**, *81*, 715–723. [CrossRef]

32. Kotwaliwale, N.; Bakane, P.; Verma, A. Changes in textural and optical properties of oyster mushroom during hot air drying. *J. Food Eng.* **2007**, *78*, 1207–1211. [CrossRef]

33. Jo, K.; Lee, J.; Jung, S. Quality characteristics of low-salt chicken sausage supplemented with a winter mushroom powder. *Korean J. Food Sci. Anim. Resour.* **2018**, *38*, 768–779.

34. Miles, P.G.; Chang, S.T. *Mushrooms: Cultivation, Nutritional Value, Medicinal Effect, and Environmental Impact*, 2nd ed.; CRC Press: Boca Raton, FL, USA, 2004; ISBN 9780849310430.

35. Kalač, P. A review of chemical composition and nutritional value of wild-growing and cultivated mushrooms. *J. Sci. Food Agric.* **2013**, *93*, 209–218. [CrossRef]

36. Dikeman, C.L.; Bauer, L.L.; Flickinger, E.A.; Fahey, G.C. Effects of stage of maturity and cooking on the chemical composition of select mushroom varieties. *J. Agric. Food Chem.* **2005**, *53*, 1130–1138. [CrossRef]

37. Yang, J.H.; Lin, H.C.; Mau, J.L. Non-volatile taste components of several commercial mushrooms. *Food Chem.* **2001**, *72*, 465–471. [CrossRef]

38. Baratzadeh, M.H.; Asoodeh, A.; Chamani, J. Antioxidant peptides obtained from goose egg white proteins by enzymatic hydrolysis. *Int. J. Food Sci. Technol.* **2013**, *48*, 1603–1609. [CrossRef]

39. Kehrer, J.P. The Haber-Weiss reaction and mechanisms of toxicity. *Toxicol.* **2000**, *149*, 43–50. [CrossRef]

40. Min-Young, K.; Philippe, S.; Joung-Kuk, A.; Jong-Jin, K.; Se-Chul, C.; Eun-Hye, K.; Su-Hyun, S.; Eun-Young, K.; Sun-Lim, K.; Yool-Jin, P.; et al. Phenolic compound concentration and antioxidant activities of edible and medicinal mushrooms from Korea. *J. Agric. Food Chem.* **2008**, *56*, 7265–7270.

41. Khan, A.A.; Gani, A.; Ahmad, M.; Masoodi, F.A.; Amin, F.; Kousar, S. Mushroom varieties found in the Himalayan regions of India: Antioxidant, antimicrobial, and antiproliferative activities. *Food Sci. Biotechnol.* **2016**, *25*, 1095–1100. [CrossRef]

42. Rahman, M.A.; Abdullah, N.; Aminudin, N. Antioxidative effects and inhibition of human low density lipoprotein oxidation in vitro of polyphenolic compounds in *Flammulina velutipes* (golden needle mushroom). *Oxid. Med. Cell. Longev.* **2015**, *2015*. [CrossRef]

43. Zeng, X.; Suwandi, J.; Fuller, J.; Doronila, A.; Ng, K. Antioxidant capacity and mineral contents of edible wild Australian mushrooms. *Food Sci. Technol. Int.* **2012**, *18*, 367–379. [CrossRef]

44. Bao, H.N.D.; Ushio, H.; Ohshima, T. Antioxidative activity and antidiscoloration efficacy of ergothioneine in mushroom (*flammulina velutipes*) extract added to beef and fish meats. *J. Agric. Food Chem.* **2008**, *56*, 10032–10040. [CrossRef]

45. Ito, H.; Ueno, H.; Kikuzaki, H. Construction of a free-form amino acid database for vegetables and mushrooms. *Integr. Food Nutr. Metab.* **2017**, *4*, 1–9. [CrossRef]

46. Ko, M.S.; Kim, S.A. Sensory and physicochemical characteristics of jeungpyun with *Pleurotus eryngii* powder. *Korean J. Food Sci. Technol.* **2007**, *39*, 194–199.

47. Han, M.; Bertram, H.C. Designing healthier comminuted meat products: Effect of dietary fibers on water distribution and texture of a fat-reduced meat model system. *Meat Sci.* **2017**, *133*, 159–165. [CrossRef]

48. Reis, F.S.; Barros, L.; Martins, A.; Ferreira, I.C.F.R. Chemical composition and nutritional value of the most widely appreciated cultivated mushrooms: An inter-species comparative study. *Food Chem. Toxicol.* **2012**, *50*, 191–197. [CrossRef]

49. Choe, J.; Lee, J.; Jo, K.; Jo, C.; Song, M.; Jung, S. Application of winter mushroom powder as an alternative to phosphates in emulsion-type sausages. *Meat Sci.* **2018**, *143*, 114–118. [CrossRef]

50. Westphalen, A.D.; Briggs, J.L.; Lonergan, S.M. Influence of muscle type on rheological properties of porcine myofibrillar protein during heat-induced gelation. *Meat Sci.* **2006**, *72*, 697–703. [CrossRef]

51. Tomasevic, I.; Tomovic, V.; Milovanovic, B.; Lorenzo, J.; Đorđević, V.; Karabasil, N.; Djekic, I. Comparison of a computer vision system vs. traditional colorimeter for color evaluation of meat products with various physical properties. *Meat Sci.* **2019**, *148*, 5–12. [CrossRef]

52. Beluhan, S.; Ranogajec, A. Chemical composition and non-volatile components of Croatian wild edible mushrooms. *Food Chem.* **2011**, *124*, 1076–1082. [CrossRef]

53. Cha, M.H.; Heo, J.Y.; Lee, C.; Lo, Y.M.; Moon, B. Quality and sensory characterization of white jelly mushroom (*Tremella Fuciformis*) as a meat substitute in pork patty formulation. *J. Food Process. Preserv.* **2014**, *38*, 2014–2019. [CrossRef]

54. Myrdal Miller, A.; Mills, K.; Wong, T.; Drescher, G.; Lee, S.M.; Sirimuangmoon, C.; Schaefer, S.; Langstaff, S.; Minor, B.; Guinard, J.X. Flavor-enhancing properties of mushrooms in meat-based dishes in which sodium has been reduced and meat has been partially substituted with mushrooms. *J. Food Sci.* **2014**, *79*, S1795–S1804. [CrossRef]

55. Ladikos, D.; Lougovois, V. Lipid oxidation in muscle foods: A review. *Food Chem.* **1990**, *35*, 295–314. [CrossRef]

56. Chauhan, P.; Pradhan, S.R.; Das, A.; Nanda, P.K.; Bandyopadhyay, S.; Das, A.K. Inhibition of lipid and protein oxidation in raw ground pork by *Terminalia arjuna* fruit extract during refrigerated storage. *Asian Australas. J. Anim. Sci.* **2019**, *32*, 265–273. [CrossRef]

57. Cossignani, L.; Giua, L.; Simonetti, M.S.; Blasi, F. Volatile compounds as indicators of conjugated and unconjugated linoleic acid thermal oxidation. *Eur. J. Lipid Sci. Technol.* **2014**, *116*, 407–412. [CrossRef]

58. Encarnacion, A.B.; Fagutao, F.; Hirayama, J.; Terayama, M.; Hirono, I.; Ohshima, T. Edible mushroom (*Flammulina velutipes*) extract inhibits melanosis in kuruma shrimp (*Marsupenaeus japonicus*). *J. Food Sci.* **2011**, *76*, C52–C58. [CrossRef]

 *foods*

Article

Drumstick (*Moringa oleifera*) Flower as an Antioxidant Dietary Fibre in Chicken Meat Nuggets

Pratap Madane [1], Arun K. Das [2,*], Mirian Pateiro [3], Pramod K. Nanda [2], Samiran Bandyopadhyay [2], Prasant Jagtap [4], Francisco J. Barba [5,*], Akshay Shewalkar [4], Banibrata Maity [4] and Jose M. Lorenzo [3,*]

[1] Division of Livestock Products Technology, ICAR-Indian Veterinary Research Institute, Bareilly 243122, India
[2] Eastern Regional Station, ICAR-Indian Veterinary Research Institute, Kolkata 700 037, India
[3] Centro Tecnológico de la Carne de Galicia, Adva. Galicia n° 4, Parque Tecnológico de Galicia, San Cibrao das Viñas, 32900 Ourense, Spain
[4] Poultry Processing Unit, Shalimar Hatcheries Limited, Grand Trunk Road, Golsi 713406, India
[5] Nutrition and Food Science Area, Preventive Medicine and Public Health, Food Science, Toxicology and Forensic Medicine Department, Universitat de València, Avda. Vicent Andrés Estellés, s/n, Burjassot, 46100 València, Spain
* Correspondence: arun.das@icar.gov.in (A.K.D.); francisco.barba@uv.es (F.J.B.); jmlorenzo@ceteca.net (J.M.L.)

Received: 9 July 2019; Accepted: 30 July 2019; Published: 1 August 2019

Abstract: The present work investigated the efficacy of Moringa flower (MF) extract to develop a functional chicken product. Three groups of cooked chicken nuggets—control (C), T1 (with 1% MF) and T2 (2% MF)—were elaborated and their physicochemical, nutritional, storage stability and sensory attributes were assessed during refrigerated storage at 4 °C up to 20 days. In addition, MF extracts were characterised in terms of chemical composition, total phenolic content and its components using high-performance liquid chromatography with a diode-array detector (HPLC-DAD), dietary fibre and antioxidant capacity. MF contained high protein (17.87 ± 0.28 dry matter), dietary fibre (36.14 ± 0.77 dry matter) and total phenolics (18.34 ± 1.16 to 19.49 ± 1.35 mg gallic acid equivalent (GAE)/g dry matter) content. The treated nuggets (T1 and T2) had significantly enhanced cooking yield, emulsion stability, ash, protein, total phenolics and dietary fibre compared to control. Incorporation of MF extract at 2% not only significantly reduced the redness/increased the lightness, but also decreased the hardness, gumminess and chewiness of the product compared to control. Moreover, the addition of MF extract significantly improved the oxidative stability and odour scores by reducing lipid oxidation during storage time. Sensory attributes of nuggets were not affected by the addition of MF extract and the products remained stable and acceptable even on 15th day of storage. These results showed that MF extract could be considered as an effective natural functional ingredient for quality improvement and reducing lipid oxidation in cooked chicken nuggets.

Keywords: lipid oxidation; natural antioxidant; sensorial properties; textural parameters; phenolic compounds; microbial analysis

---

1. Introduction

Muscle foods are an excellent source of high-quality protein with high biological value. Moreover, the bioavailability of micronutrients such as iron, selenium, vitamins A, B12, folic acid, sodium, potassium and magnesium of these matrices is much higher than from plant sources [1]. In spite of being nutritious, meat is deficient in dietary fibre, a complex mixture of polysaccharides, which is naturally present as a part of plant material in cereals, vegetables, fruits and nuts. Lack of adequate quantity of dietary fibres in our diet has been involved in several health disorders such as colon cancer, obesity and cardiovascular diseases [2]. On the other hand, a fibre-rich diet is lower in energy density and richer in micronutrients thus reducing several disorders [3], and thereby promotes a healthier lifestyle. However, many food and food products, including meat products, not only lack of minimum amounts of dietary fibre [4] to fulfil the recommended dietary intake, but also differ in the quantity and composition of fibres.

As per the American Dietetic Association, the recommended dietary fibre intake for a healthy adult should be approximately 25 to 30 g/day, being the insoluble/soluble fibre ratio 3:1. Hence, to get the above health benefits of dietary fibre, attempts have been made to incorporate dietary fibres from various sources in meat product formulation to promote various technological benefits and the acceptability of meat products benefitting human health [5].

In a developing country like India, industrialisation, rapid urbanisation, globalisation as well as increasing number of women workforce are factors contributing to a rapidly growing inclination towards fast and convenient meat and food products [6]. However, meat and meat products contain an important amount of unsaturated fatty acids, which are sensitive to lipid oxidation [7,8]. Further, processing such as grinding, chopping, flaking, emulsification and cooking also accelerate lipid oxidation of meat and meat products [8]. Moreover, most meat products are processed using vegetable oils in order to overcome the problems of saturated fatty acids and cholesterol associated with animal fats [9,10]. These vegetable oils, with a high degree of polyunsaturation, accelerate oxidative processes, leading to meat quality deterioration. Therefore, lipid oxidation is considered the major causes of meat quality deterioration with development of undesirable flavours and odours, thus reducing the nutritional, sensorial and functional properties of meat products as well as consumer acceptability [11,12].

The use of antioxidants is considered as an effective method to minimise or inhibit lipid oxidation as well as inhibit the formation of toxic oxidation products in muscle foods, thereby improving the shelf-life of products [13–15]. Although synthetic antioxidants have been widely used in the meat industry, consumer concerns over safety issues of products have renewed the interest of food industry in search of antioxidants from natural sources [16]. Thus, both dietary fibre and natural antioxidants are considered as two important dietary fractions and could be very valuable in improving meat product quality and storage stability.

The natural ingredients having dual properties, i.e., a source of dietary fibre besides having antioxidant potential are known as antioxidant dietary fibres (ADF). ADF is defined as dietary fibre concentrate containing a significant content of natural antioxidants linked to the dietary fibre (DF) matrix in a single material [17]. To counter the shortcomings (low fibre content and oxidation) associated with meat and meat products, researchers and meat processors are continuously searching for natural ingredients and materials/extracts, especially of plant origin, to use as additives. The incorporation of ADF in meat products not only increases the shelf life during storage by inhibiting lipid oxidation due to presence of phenolic antioxidants, but also improve the texture, physicochemical and sensory properties of meat products [4,18,19].

Moringa oleifera, commonly known as horse radish tree or drumstick tree, is one of the most widely cultivated species native to the sub-Himalayan tracts of India. Almost all parts of this plant, like fruit (pods), gum, root, seed, bark, leaf, flowers and seed oil, are used as a nutritional and nutraceutical resources for human and animal diets [20]. The leaves and flowers are good source of protein and dietary fibre with an adequate profile of amino acids and ash [21,22]. The extracts of flowers of *M. concanensis* (fresh or dried) contain a great amount of ascorbic acid, polyphenols, tannins and flavonoids with high DPPH scavenging activity [23]. Considering the benefits of both dietary fibre and antioxidants in a single material, the objective of this study was to assess the potential use of Moringa flower (*M. oleifera*) as ADFs or functional ingredients in meat food system to enhance the nutritional quality, storage stability and acceptability of meat products.

2. Materials and Methods

2.1. Reagents and Plant Materials

Fresh mature moringa flowers (*M. oleifera*) were collected from a local market of Kolkata, India. Flowers were cleaned thoroughly to remove extraneous dirt, dried completely in a hot air oven at 45 ± 2 °C, ground in a grinder (Kenstar, Mumbai, India) and sieved (#60 mesh sieves). The powder obtained was stored in an air tight container at room temperature until further use. Chicken meat (breast) was obtained from West Bengal Livestock Development Corporation, Kolkata, India and kept under frozen storage at −18 °C till further processing. Different chemicals such as methanol, butylated hydroxytoluene (BHT), α-amylase, protease, amyloglucosidase trichloroacetic acid, sodium carbonate, 2-thiobarbituric acid, 2,2-diphenyl-1-picrylhydrazyl (DPPH), phosphate buffer, potassium ferricyanide, F-C reagents and gallic acid, caffeic acid, ferulic acid and quercertin were purchased from Sigma-Aldrich (Mumbari, India). All the chemicals used for this experiment were of analytical grade.

2.2. Preparation of Moringa Flower Extract

Moringa flower (MF) was extracted using either aqueous (AE) or aqueous ethanol (AEH) as solvent (60:40, *v/v*). Briefly, for preparation of the extracts, 2 g of each MF was accurately weighed into separate conical flasks. To this, 100 mL of solvent were added and the whole content was held at room temperature (27 ± 1 °C) for 10 h, stirring frequently with a glass rod. The mixture was shaken at constant rate (500 rpm) using a shaker, vortexed at high speed for 10 min, and finally centrifuged (REMI NEYA 8, Kolkata, India) at 5000x rpm for 10 min. The content of each extract was then passed through Whatman filter paper No. 1 (HiMedia®, Mumbai, India). The resulting extract was kept in a container and stored at 2 °C for further studies. The extracts (aqueous or aqueous ethanol) obtained from different solvent were analysed for total phenolic contents (TPC), 1, 1 diphenyl-2-picrylhydrazil (DPPH) radicals scavenging activity and ferric reducing antioxidant power (FRAP) assays. The efficacy of the extracts was determined based on the weight of respective dry powders.

2.3. Analysis of Polyphenols and Antioxidant Capacity

The total phenolics content of the extracts was assessed using the Folin–Ciocalteu (F-C) method previously described by Singleton et al. [24], while in cooked nuggets the procedure described by Escarpa & González [25] with slight modifications was used. The results were expressed as mg gallic acid equivalents (GAE) /g of dry matter.

HPLC-DAD phenolic composition of MF extracts was carried out according the method of Zeb [26], using an Agilent 1260 Infinity HPLC system equipped with a quaternary pump, degasser, autosampler and diode-array (DAD) detector. An Agilent rapid resolution Zorbax Eclipse plus C18 (4.6 × 100 mm, 3.5 μm) column was used for the separation, using a gradient elution at 0.6 mL/min with gradient program (0–20 min, 95–75% A; 20–40 min, 75–50% A; 50–20%, 40–50 min A; 50–60 min, 20% A) with 1% formic acid in water as solvent A and methanol as solvent B. The chromatograms were obtained at 280 nm.

On the other hand, the method proposed by Fargere et al. [27] was used to determine the free radical scavenging activity (FRSA) using DPPH assay. Free radical scavenging activity (FRSA) was calculated using the following formula.

$$FRSA(\%) = \frac{(absorbance\ control - absorbance\ sample)}{absorbance\ control} \times 100 \qquad (1)$$

The Ferric reducing antioxidant power (FRAP) assay was assessed following the procedure previously established by Oyaizu [28].

2.4. Determination of Dietary Fibre Content

The dietary fibre composition was analysed according to the enzymatic–gravimetric method [29]. A phosphate buffer was used to disperse the sample and a sequential enzymatic digestion using α-amylase, protease and amyloglucosidase was carried out. The total dietary fibre (TDF) was calculated as the sum of insoluble dietary fibre (IDF) and soluble dietary fibre (SDF) obtained from the enzymatic digestion.

2.5. Preparation of Chicken Nuggets

Approximately 8 kg of meat sample (chicken breast meat, breast trimmings and chicken skin), obtained from West Bengal Livestock Development Corporation (West Bengal, India), was minced twice (minced through a 6 mm grinding plate followed by 4 mm plate) in a meat mincer. After mincing, meat samples were divided into three (3) different batches. The composition of ingredients for each batch was based on the standard formulation (Table 1). The first batch was considered as control (meat without any ADF), whereas MF at 1% and 2% level were used as ADF for chicken nuggets in formulations (T1 and T2) replacing equal per cent of breast trimmings.

Table 1. Formulation of chicken nuggets with different levels of Moringa flower (MF) as antioxidant dietary fibre.

Ingredients (%)	Treatment		
	Control	T1	T2
Chicken meat	32.0	32.0	32.0
Breast trimming	20.0	19.0	18.0
Chicken skin	10.0	10.0	10.0
Ice flakes	20.0	20.0	20.0
Refined vegetable oil	8.00	8.00	8.00
Salt	1.50	1.50	1.50
Condiments *	4.00	4.00	4.00
Refined wheat flour	2.40	2.40	2.40
Dry spice mix **	1.80	1.80	1.80
Sodium nitrite (ppm)	150	150	150
Poly-phosphate	0.30	0.30	0.30
ADF powder %	0.0	1.0	2.0

Treatments: Control = no additive; T1 = 1.0% Moringa flower (MF) extract and T2 = 2.0% Moringa flower extract. * Condiments: garlic and onion (4:1). ADF = Antioxidant dietary fibre. ** Dry spice mix (18 g/kg nuggets)—aniseed, black pepper, capsicum, caraway seed, cardamom, cinnamon, cloves, coriander powder, cumin seed, turmeric and dried ginger.

All batches of minced meat samples were prepared separately in a bowl chopper. In order to prepare meat emulsion, salt, phosphate and nitrite were added and mixed to the meat using a bowl chopper. Ice flakes were added to keep temperature at ~8 ± 2 °C. Afterwards, condiments, dry spice mix, and fine wheat flour were added, and chopping was continued till uniform mixing of all the ingredients. The emulsion prepared was steam-cooked (100 °C) for 40 min to get cooked chicken meat nugget. The blocks were sliced uniformly to obtain small cubes (size) of chicken nuggets. The formulated nuggets (C, T1 and T2) were aerobically packaged in low density polyethylene (LDPE) pouches and kept under refrigerated conditions (4 ± 1 °C) to evaluate different physicochemical parameters, including storage stability (0, 5, 10, 15 and 20 days) and sensory attributes. The whole experiment was conducted three times and samples were analysed in duplicate.

2.6. Microbial Analysis

Total plate count (cfu/g) was determined by using pour plate method according to Lorenzo et al. [30]. Ten grams of meat sample were homogenised in 90 mL of sterile peptone water (0.1%). Appropriate serial dilutions were prepared in 0.1% sterile peptone water and plated in duplicate with plate count agar, incubated at 37 °C for 48 h. Microbial colonies from the plates were counted and expressed as log cfu/g.

2.7. Physicochemical Analysis

Moisture, protein, fat and ash content of the cooked meat nugget samples were determined by the procedures previously described by Association of Official Analytical Chemists AOAC [31]. The pH of sample was determined by combination electrode digital pH meter (Benchtop pH meter, BR Biochem, PHS-25CW, New Delhi, India). Briefly, 10 g of sample was homogenised with the help of tissue homogeniser (Omni International, Kennesaw GA, United States) for approximately a minute in 50 mL of distilled water. The homogenised sample was kept for 5 minutes and then mixed again by shaking rod. The pH value was recorded by immersing the electrode directly into the suspension.

The colour profile of thick slice of chicken nugget or fresh chicken block was measured using Hunter Color Lab (Mini XE, Portable, HunterLab, Reston, VA, United States) to record Hunter L*, a*, and b* values. The instrument was calibrated using light trap/black glass and white tile provided with the instrument. The instrument was directly put on the surface of meat block and reading was taken at three different points. The textural properties of nuggets were evaluated using a texture analyser (Stable Micro System, Model TA. HDi, Surrey, UK). Texture profile analysis (TPA) was performed using central cores of five pieces of each sample (2 cm × 2 cm × 2 cm), which were compressed twice to 80% of the original height by a compression probe (P 75). A crosshead speed of 2 mm/s was used. The following parameters were determined; hardness (N/cm^2), springiness (cm), cohesiveness, gumminess (N/cm^2) and chewiness (N/cm).

Cooking yield of nuggets was determined by recording the weight of each meat block before and after cooking and expressed as percentage according to the following equation.

$$Cooking\ yield\ (\%) = \frac{(wt\ of\ cooked\ meat\ block)}{wt\ of\ raw\ meat\ block} \times 100 \tag{2}$$

Expressible water was estimated using the method of Das et al. [32] with slight modifications. Approximately 5 g of cooked sample was weighed and placed on 2 layers of Whatman no. 1 filter paper. The sample was placed at the bottom of 50 mL centrifuge tube and centrifuged at 1500 g (Remi, Mumbai, India) for 15 min. Immediately after centrifugation, meat samples were reweighed and the percentage of expressible water was calculated according to the following equation.

$$Expressible\ water\ (\%) = \frac{(initial\ weight - final\ weight)}{initial\ weight} \times 100 \tag{3}$$

2.8. Thiobarbituric Acid Reacting Substances (TBARS) Values

The evaluation of lipid stability was performed by measuring TBARS values following the method proposed by Witte et al. [33]. The TBARS value was calculated as mg malonaldehyde (MDA) per kg of the sample by multiplying the absorbance value with a factor of 5.2.

2.9. Sensory Evaluation of Chicken Nuggets

Sensory evaluation of the control and treated chicken nuggets was used for various sensory attributes by trained and experienced panellists familiar with the characteristics of the meat product. Just before sensory evaluation, nugget samples (C, T1 and T2) were warmed in a microwave oven for 20 s, coded and then served to evaluate for general appearance, colour, flavour, binding, texture, juiciness and overall acceptability using 8-point descriptive scale [34], where 8 = extremely desirable and 1 = extremely undesirable. Plain potable water was provided to rinse the mouth in between the samples.

2.10. Statistical Analysis

The present study was repeated three times and, in each replication, measurements of all parameters were done in duplicate. The analysis was carried out using statistical package for the social sciences (SPSS) software (version 20.0, IBM SPSS, Armonk, NY, United States) and for storage study, data was analysed using two-way analysis of variance (ANOVA) with interaction using SPSS software where treatment (control, T1 and T2) and storage time (0, 5, 10, 15 and 20) were considered as factors. The experiment was carried out in 3 × 5 factorial design according to a completely randomised design. The obtained data were subjected to variance analysis, and Duncan's multiple range test was used for comparing the means to find out the effect of antioxidant dietary fibres on various parameters and the differences were considered at $\alpha = 0.05$ level. The values were presented as mean along with standard error (mean ± standard error).

3. Results and Discussion

3.1. Chemical Composition, Phenolic Content and Antioxidant Activity of MF Extract

The chemical composition and dietary fibre contents of the MF extract are shown in Table 2. The protein, ash and lipid content of MF were 17.87 ± 0.28%, 7.87 ± 0.45% and 2.95 ± 0.07%, respectively. In general, the chemical composition depends on the edible part of a plant being analysed. Our findings are in disagreement with the data reported by other authors in MF extract [35,36]. This fact could be attributed to the soil type, cultivars, stage of maturity of flowers and influence of the climatic or weather conditions in the region [37]. As presented in Table 2, MF extract had 36.14 ± 0.77% TDF, 3.90 ± 0.14% SDF and 32.24 ± 0.82% IDF contents. The content of total dietary fibre in MF extract is similar to that reported by Sánchez-Machado et al. [36]. The presence of great amount of IDF and SDF in the MF extract indicate that MF is a promising source of dietary fibre with a very good physiological effect; better than some cereals like wheat bran (2.9% SDF; 41.1% IDF), and oat bran (3.6% SDF; 20.2% IDF) as reported by Grigelmo-Miguel et al. [38].

Table 2. Proximate composition, total phenolic content (TPC), IC_{50} (μg/mL) and phenolic compounds of Moringa flower (mean values ± error standard of six samples).

Proximate composition (g/100 g dry matter)	
Protein	17.87 ± 0.28
Lipid	2.95 ± 0.07
Ash	7.87 ± 0.45
Total dietary fibre (TDF)	36.14 ± 0.77
Soluble dietary fibre (SDF)	3.90 ± 0.14
Insoluble dietary fibre (IDF)	32.24 ± 0.82
Nonstructural carbohydrates (NSC)	35.17 ± 1.25
Antioxidant capacity	
TPC (mg GAE/g dry matter) from AE	18.34 ± 1.16
TPC (mg GAE/g dry matter) from AEH	19.49 ± 1.35
IC_{50} μg/mL from aqueous extract	126.20 ± 1.45
IC_{50} μg/mL from aqueous ethanol extract	121.42 ± 1.28
Phenolic compounds (mg/kg dry matter)	
Caffeic acid	ND
Ferulic acid	270.08 ± 3.78
Quercertin	15.14 ± 0.40
Gallic acid	ND

AE = Aqueous extract; AEH = Aqueous ethanolic extract, ND = Not detected, IC = Inhibitory concentration.

In the present study, the aqueous ethanolic extract of MF presented the higher total phenolic compounds (TPC) (19.49 ± 1.35 mg GAE/g powder) followed by water extract (18.34 ± 1.16 mg GAE/g powder), although nonsignificant (Table 2). The higher TPC might be attributed to different degree of polarity of the solvents used for the extraction of polyphenolic compounds, and thus could have contributed significantly to the antioxidant and free radical scavenging activity. Similar observation was also demonstrated by Tekle et al. [39], who reported that the total polyphenolic content of MF extract was relatively higher in ethanolic extract compared to its water counterpart.

The phenolic compounds in MF extracts were identified and quantified as shown in Table 2. Among the phenolics, ferulic acid was present in high concentration and other compounds, including flavonoids (quercetin) were also present. In fact, antioxidant activities of plants mainly depend on the amount of phenolic acids (gallic acid, feluric acid and caffeic acid) and flavonoids (catechin, myricetin and quercetin). Several studies indicate that flavonoids are the main contributor for plant's antioxidant activity [40,41]. However, different phenolics exhibit different antioxidant potential, and other compounds may influence the antioxidant activities [42]. The strong antioxidant activity exhibited by MF extracts could be due to abundance of ferulic acid and quercetin.

According to DPPH scavenging activity, our results showed that aqueous ethanol extract of MF contained higher radical scavenging activity than aqueous extract (Figure 1). Both extracts demonstrated purple bleaching reaction at increased concentrations, showing the presence of compounds responsible as free radical scavengers thus reducing the initial DPPH concentration.

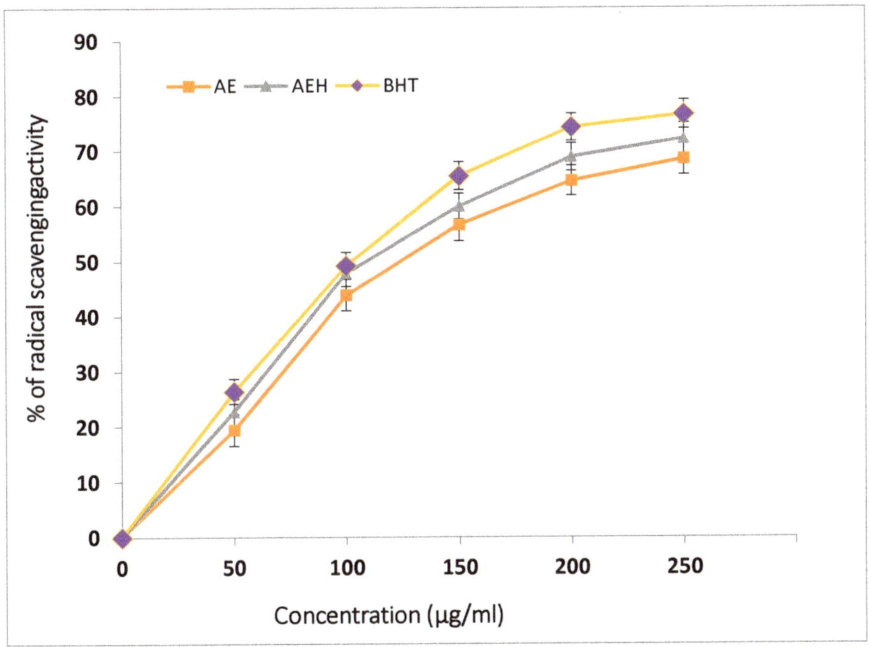

Figure 1. DPPH radical scavenging activity of MF extracts. AE = Aqueous extract; AEH = Aqueous ethanolic extract; BHT = Butylated hydroxytoluene.

Figure 1 clearly illustrates that the radical scavenging activity of Moringa flowers in both extracts was comparable to butylated hydroxytoluene (BHT) at concentrations from 0 to 250 µg/mL. The levels of inhibition of DPPH radical by the AE, AEH and BHT were 68.52, 72.22 and 76.64%, respectively, and in a concentration-dependent manner. The BHT used as positive control displayed similar level of inhibition at 250 µg/mL, which was nonsignificantly different in comparison to AE. A high correlation exists between DPPH radical scavenging potential and TPC of plant extracts. A study conducted by Siddhuraju et al. [43] reported that at a dosage ranging from 0.2 to 0.6 mg of acetone extract, *M. oleifera* (pericarp of immature drumstick and flower) and *S. grandiflora* (flower and leaf) showed higher free radical scavenging activity (8.45–65.03%). Sreelatha and Padma [44] noticed that methanol extract of *M. oleifera* leaves significantly reduced DPPH radicals, though lower than our observed results. These variations could be due to difference in the polarity of solvents and geographical location of the plants [44].

In terms of IC_{50}, the lowest value was shown by the positive control, BHT (118 µg/mL), followed by AEH (121.42 µg/mL) and AE (126.20 µg/mL) of MF extracts (Table 2). Alhakmani et al. [45] reported that DPPH radical scavenging activity of *M. oleifera* flower extract was compared with standard ascorbic acid. Although standard antioxidant had higher scavenging activity at all tested concentrations, the flower extract still showed good free radical scavenging activity. This radical scavenging activity of extracts might be related to the nature of phenolics, thus contributing to their electron transfer/hydrogen donating ability [22,43].

Regarding FRAP assay, MF extract presented good activity on a dose dependent manner. As the aqueous ethanol-extracted MF extract exhibited an overall higher activity at the highest concentrations compared to water extract MF (Figure 2). This outcome showed that it was more efficient in extracting antioxidants from plant materials. In this regard, Tekle et al. [46] also found that ethanolic extract of M. oleifera leaf and flower exhibiting higher ferric reducing antioxidant activity ($p < 0.05$) compared to the synthetic antioxidant, BHT.

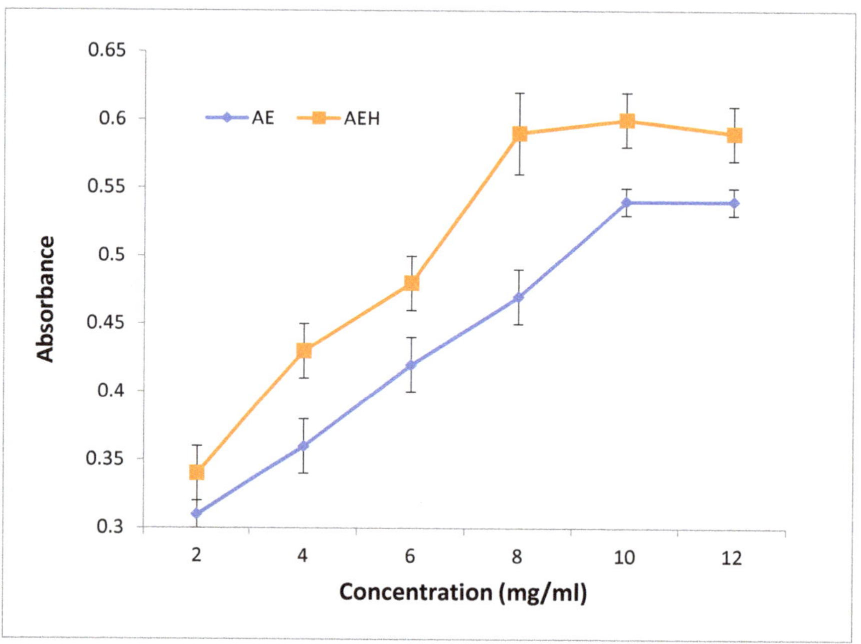

Figure 2. Ferric reducing antioxidant power of MF extracts. AE = Aqueous extract; AEH = Aqueous ethanolic extract.

3.2. Effect of MF Extract on Physico-Chemical Properties of Meat Nuggets

Table 3 shows the influence of the MF extract on the physicochemical properties of meat nuggets. There was a significant ($p < 0.05$) decrease in emulsion pH with incorporation of MF extract at both levels used in this study. The highest pH value was observed in control (6.33) followed by T1 (6.25) and T2 (6.22) batches, although nonsignificant ($p > 0.05$). The decrease in pH of meat emulsion might be due to acidic pH value (5.44) of MF extract. Our findings agree with the results of Devatkal et al. [47], who noticed that kinnow rind extract significantly decreased the pH values of cooked goat meat patties due to its acidic pH. In a similar study, Habib et al. [48] also found a decrease in pH on the incorporation of pomegranate rind powder at three different levels. However, Das et al. [49] noticed that the use of *M. oleifera* leaves extract did not modify the pH of raw and cooked goat meat patties, whereas Hazra et al. [50] observed that the addition of *M. oleifera* leaves extract increased ($p < 0.05$) the pH values of cooked ground buffalo meat.

Table 3. Effect of Moringa flower on physico-chemical parameters of chicken nuggets (mean values ± error standard of six samples).

Parameters	Treatments			Sig.
	Control	T1	T2	
Emulsion pH	6.33 ± 0.02 [a]	6.25 ± 0.02 [b]	6.22 ± 0.02 [b]	**
Emulsion Stability (%)	94.45 ± 0.10 [c]	95.56 ± 0.09 [b]	96.47 ± 0.29 [a]	**
Cooking yield (%)	96.79 ± 0.07 [b]	97.26 ± 0.09 [a]	97.83 ± 0.22 [a]	*
Total phenolic content (mg GAE/g)	0.059 ± 0.02 [c]	0.789 ± 0.09 [b]	1.121 ± 0.15 [a]	***
Expressible water (%)	27.14 ± 1.17	24.42 ± 3.00	21.71 ± 2.05	ns
Chemical composition (g/100 g)				
Moisture	67.29 ± 0.54	66.36 ± 0.82	65.74 ± 0.56	ns
Protein	14.38 ± 0.34 [a]	15.27 ± 0.29 [ab]	16.32 ± 0.66 [b]	**
Fat	13.76 ± 0.49	14.06 ± 0.35	14.69 ± 0.62	ns
Ash	2.37 ± 0.49 [a]	2.64 ± 0.13 [ab]	2.91 ± 0.89 [b]	*
Total dietary fibre	0.76 ± 0.03 [a]	1.39 ± 0.04 [b]	2.03 ± 0.06 [c]	***
Textural parameters				
Hardness (N/cm^2)	69.71 ± 2.43	65.52 ± 3.89	64.33 ± 6.49	ns
Springiness (cm)	0.67 ± 0.05	0.65 ± 0.02	0.63 ± 0.02	ns
Cohesiveness	0.30 ± 0.02	0.33 ± 0.01	0.34 ± 0.01	ns
Gumminess (N/cm^2)	20.79 ± 0.02	20.01 ± 1.44	18.13 ± 3.04	ns
Chewiness (N/cm)	14.25 ± 2.25	13.79 ± 1.63	11.95 ± 2.54	ns

Treatments: Control = no additive; T1 = 1.0% Moringa flower extract and T2 = 2.0% Moringa flower extract. [a–c] Mean values in the same row not followed by a common letter differ significantly. Sig. Significance; ns: not significant; * $p < 0.05$; ** $p < 0.01$; *** $p < 0.001$.

Statistical analysis showed that the moisture and lipid content did not differ significantly ($p < 0.05$) between control and treated samples (Table 3). Our findings agree with the data reported by Al-Juhaimi et al. [51], who observed that the moisture content of raw patties decreased as the percentage of *Moringa oleiferi* seed flour increased, but the decline rate was found to be not significant. In addition, Hazra et al. [50] also found that the inclusion of *Moringa oleiferi* leaf extract did not modify the moisture and lipid content of cooked ground buffalo meat. The protein content of the control nugget was 14.38%, whereas the treated nuggets (T1 and T2 groups) presented protein values of 15.27% and 16.32%, respectively. In fact, Moringa flowers have a very high crude protein content varying between 18.92 and 26.16% [35,36]. Therefore, the increase in protein percentage of chicken nuggets might be due to higher protein (17.87%) percentage of MF extract as found in this study. There was a significant difference in the percentage of ash content between control and treated nuggets. The increased ash content in treated chicken nuggets might be due to high ash content of MF (7.87%). A similar trend was reported by Al-Juhaimi et al. [51] who noticed that the ash content of raw patties increased as the percentage of *Moringa oleifera* seed flour increased in cooked ground buffalo meat.

The incorporation of MF significantly ($p < 0.05$) enhanced the emulsion stability of treated nuggets than the control (Table 3). The control and treated chicken nuggets (T1 and T2 groups) showed a 94.45%, 95.56% and 96.47% emulsion stability, respectively. The probable reason for increased emulsion stability could be due to the presence of dietary fibre in MF (36.14%). In this regard, Das et al. [18] also observed that bael pulp residue at 0.5% significantly improved ($p < 0.05$) the emulsion stability in goat meat nuggets. Similarly, Sarıçoban et al. [52] studied the effect of different concentrations (2.5%, 5%,

7.5% and 10%) of lemon albedo (raw and dehydrated) on the functional properties of emulsions and found an increase in the emulsion capacity at 5 percent of added emulsions. A similar trend was also observed in emulsion stability values. Malav et al. [53] reported that there was an increasing trend in emulsion stability with increase in levels of red kidney bean powder. The emulsion stability and cooking yield increased on incorporation of dietary fibre extracted from rice bran at different levels in meat batters [54].

On the other hand, a significant ($p < 0.05$) improvement in cooking yield was observed due to the incorporation of MF at different levels (Table 3). The higher percentage of cooking yield (97.83% and 97.26%) was observed in nuggets with MF than the control (96.79%). The articles available in the literature regarding this aspect indicate that non-meat ingredients with high dietary fibre content, when used in emulsion type of meat products, improve the cooking yield. A similar trend was reported by Ham et al. [55] who noticed that increasing lotus rhizome powder levels as a source of ADF lowered the cooking loss of emulsion sausages (5.89–6.25%) significantly ($p < 0.05$) than that of control sausage (7.31%). In addition, Anderson and Berry [56] also observed that 10% fat beef patties extended with pea fibre had high cooking yield. The probable reasons for the increased cooking yield of chicken nuggets with MF extract could mainly be attributed to the presence of high amount of dietary fibre and its ability to bind more water and fat as reported by Verma and Banerjee [4]. However, this finding disagrees with the data reported by Das et al. [49] who observed that the use of *M. oleifera* leaves extract did not influence the cooking loss of goat meat patties.

On the contrary, Hazra et al. [50] found that the inclusion of *M. oleifera* leaves extract showed a significant reduction in cooking loss of cooked ground buffalo meat. In this study, the expressible water content of nuggets ranged between 21.71% and 27.14% (Table 3). Although chicken nuggets with MF extract had lower expressible water, indicating higher water-holding capacity (WHC), there was no significant effect ($p > 0.05$) compared to control group. This result agrees with those reported by Vural et al. [57] who observed that the use of sugar beet fibre increased the water-holding capacity of frankfurter without any significant changes on sensory properties. In addition, Chang and Carpenter [58] also reported that more water was retained in frankfurters with an increase in oat bran level. Our findings clearly indicate that fibre can be used in cooked meat products to increase the WHC.

Addition of MF significantly ($p < 0.001$) increased the total dietary fibre (TDF) and total phenolics content (TPC) in chicken nuggets (Table 3). The TDF content was the highest in nuggets with 2% MF extract (2.03%) followed by T1 group (1.39%), while the lowest values were found for control samples (0.76%). This outcome agrees with the data reported by Verna et al. [6] who noticed that the incorporation of guava powder as ADF significantly improved the TDF content of meat nuggets. Similarly, TPC content was significantly higher in treated nuggets (0.789 and 1.121 mg GAE/g for T1 and T2 groups, respectively) than control samples (0.059 mg GAE/g). Such a high dietary fibre and TPC level in treated chicken nuggets might be due to use of MF extract, which had very high phenolic content (36.14 mg/g dry powder) and good source of dietary fibre (36.14%). This result agrees with the data observed by Das et al. [49] who showed that TPC of cooked goat meat patties with *M. oleifera* leaves extract was significantly ($p < 0.05$) higher compared to control group.

In addition, Das et al. [18] also found significantly increased ($p < 0.05$) TPC by incorporating bael pulp residue as ADF in meat products. Moreover, Das et al. [34] reported that sheep meat nuggets incorporated with 1% and 1.5% litchi fruit pericarp extract had significantly higher TPC than control nugget.

The incorporation of MF extract did not influence ($p > 0.05$) the textural parameters (hardness, cohesiveness, gumminess and chewiness) of the product, although slightly lower hardness values were found in T2 group, indicating softer chicken nuggets compared to control (Table 3). A similar trend was found by Verma et al. [6] who observed that hardness, adhesiveness, cohesiveness, gumminess and chewiness values were not significantly affected ($p > 0.05$) by the addition of guava powder. Similarly, Choi et al. [54] studied the addition of rice bran fibre on the textural properties of heat-induced gel, and found that not only hardness, but springiness, cohesiveness, gumminess and chewiness were also lower in samples with added rice bran fibre relative compared to control treatment. Moreover, Ham et al. [55] reported that textural properties, notably hardness, cohesiveness, and gumminess of cooked sausages were unaffected ($p > 0.05$) even when formulated with different levels of lotus rhizome powder. Other textural properties recorded in this study were nonsignificantly different ($p > 0.05$) between control and treatment chicken nuggets, though all the values decreased with increasing levels of MF. Wan Rosli et al. [39] while studying the textural properties of chicken patties formulated with different levels (0, 25% or 50%) of grey oyster mushroom, as a source of fibre and fat replacer, found lower cohesiveness, gumminess and chewiness values compared to control.

3.3. Effect of MF on pH, TBARS Values and Total Plate Count of Chicken Nuggets during the Storage Time

The pH values of the control and treated chicken nuggets, which were aerobically packaged and stored under refrigerated conditions, were evaluated at five-day intervals up to 20th day and are presented in Table 4. Statistical analysis displayed significant difference ($p < 0.05$) between storage days and treatments. As storage period progressed, a significant ($p < 0.05$) increase in pH values was observed in both the control and treated nuggets. The pH values increased from 6.30 ± 0.02 to 6.50 ± 0.01, from 6.27 ± 0.01 to 6.36 ± 0.01 and from 6.26 ± 0.01 to 6.37 ± 0.01 from day 0 to day 20 for control, T1 and T2, respectively. The increase in pH during the storage period of meat product might be because of accumulation of metabolites due to the growth of Gram-negative bacteria like Pseudomonas, Moraxella, Acinetobacter, etc. [59]. Das et al. [60] also observed an increase in pH of ground and cooked meat added with curry leaf (*Murraya koenigii*) during the storage time. However, the increase in pH during the refrigerated period was significantly ($p < 0.05$) less in treated samples compared to control treatment, which might be due to the inhibitory effects of MF extract on oxidation of protein and lipid and some antimicrobial effects of the plant powder [61].

Table 4. Effect of Moringa flower on pH, thiobarbituric acid reactive substances (TBARS) and microbial count (log cfu/g) of chicken nuggets during the storage time (mean values ± error standard of six samples).

Treatment	Storage Time (Days)					Sig.
	0	5	10	15	20	
pH						
Control	6.30 ± 0.02 [d]	6.34 ± 0.03 [cd]	6.39 ± 0.03 [bcx]	6.45 ± 0.04 [abx]	6.50 ± 0.01 [ax]	***
T1	6.27 ± 0.01 [c]	6.29 ± 0.01 [bc]	6.32 ± 0.01 [aby]	6.33 ± 0.01 [aby]	6.36 ± 0.01 [ay]	***
T2	6.26 ± 0.01 [d]	6.30 ± 0.02 [cd]	6.32 ± 0.01 [bcy]	6.36 ± 0.01 [aby]	6.37 ± 0.01 [ay]	***
Sig.	ns	ns	***	***	***	
TBARS (mg malonaldehyde/kg of sample)						
Control	0.37 ± 0.01 [e]	0.53 ± 0.01 [dx]	0.89 ± 0.01 [cx]	1.38 ± 0.02 [bx]	1.94 ± 0.05 [ax]	***
T1	0.36 ± 0.01 [d]	0.38 ± 0.01 [dy]	0.45 ± 0.01 [cy]	0.58 ± 0.01 [by]	0.84 ± 0.02 [ay]	***
T2	0.36 ± 0.01 [d]	0.37 ± 0.02 [dy]	0.42 ± 0.01 [cy]	0.52 ± 0.01 [bz]	0.81 ± 0.01 [ay]	***
Sig.	ns	***	***	***	***	
Total Plate Count (log cfu/g)						
Control	2.74 ± 0.06 [e]	4.10 ± 0.07 [dx]	5.13 ± 0.06 [cx]	6.12 ± 0.08 [bx]	6.46 ± 0.04 [ax]	***
T1	2.64 ± 0.09 [d]	3.37 ± 0.06 [cy]	3.84 ± 0.07 [by]	4.10 ± 0.05 [by]	4.66 ± 0.10 [ay]	***
T2	2.71 ± 0.04 [d]	3.52 ± 0.12 [cy]	3.83 ± 0.10 [by]	4.02 ± 0.06 [by]	4.51 ± 0.05 [ay]	***
Sig.	ns	***	***	***	***	

Treatments: Control = no additive; T1 = 1.0% Moringa flower extract and T2 = 2.0% Moringa flower extract. [a–e] Mean values in the same row not followed by a common letter differ significantly among storage times. [x–y] Mean values in the same column not followed by a common letter differ significantly among treatments. Sig. Significance; ns: not significant; *** $p < 0.001$.

The mean ± SE values of total plate count of aerobically packaged control and treated chicken nuggets (T1 and T2) during refrigerated storage up to 20 days is presented in Table 4. During storage, a significant increase ($p < 0.05$) in microbial count was observed in control and treated products (T1 and T2) at each interval of storage period except on 0 day, where the counts were comparable. Total plate count increased from 2.74 ± 0.06 to 6.46 ± 0.04 $\log_{10}$ cfu/g, from 2.64 ± 0.09 to 4.66 ± 0.10 $\log_{10}$ cfu/g and from 2.71 ± 0.04 to 4.51 ± 0.05 $\log_{10}$ cfu/g, from day 0 to day 20 for control, T1 and T2 groups, respectively. An increase of total plate count of chicken sausage during refrigerated storage was noticed by Sallam et al. [62]. However, at the end of storage time, the total plate count of treated chicken nuggets was significantly lower compared to the control group (6.46 vs. 4.66 and 4.51; $p < 0.001$ for control, T1 and T2 groups, respectively). This result might be due to its richness in polyphenolic compounds [45] exerting antimicrobial [61] effects.

It has been well documented by some researchers that polyphenols from MF extract have microbial activities against a number of pathogenic bacteria [61]. The effectiveness of MF extract in lowering the total plate count of chicken nuggets are in agreement with the previous findings reported by Das et al. [18] who observed that incorporation of bael pulp residue as ADF was very much effective ($p < 0.05$) in controlling the microbial counts in goat meat nuggets throughout the 20 days of storage period.

The TBARS values of aerobically packaged chicken nuggets studied at regular intervals up to 20 days under refrigerated storage conditions is presented in Table 4. The TBARS values of chicken nuggets, irrespective of treatments, increased significantly ($p < 0.05$) from 0.37 ± 0.01 to 1.94 ± 0.05 mg MDA/kg, from 0.36 ± 0.01 to 0.84 ± 0.02 mg MDA/kg and from 0.36 ± 0.01 to 0.81 ± 0.01 mg MDA/kg, from day 0 to day 20 for control, T1 and T2 treatments, respectively. This increase in TBARS value throughout the storage period might be due to the lipid oxidation and production of volatile metabolites in presence of oxygen during aerobic storage [63–66]. Although TBARS value of all treated groups increased, but the rate of increase was comparatively slower in case of treated nuggets indicating more oxidative stability due to the presence ADF from MF extract. As observed on day 15, treated nuggets were within the spoilage limit, whereas during the same period, control nuggets (1.38) crossed the acceptable limit of 1 mg MDA/kg. The polyphenolic compounds present in the MF extract may be the reason for the strong antioxidant ability as reported by different researchers [22,47].

While comparing the treated groups, it was found that chicken nuggets in T2 group retarded the oxidation process more efficiently by maintaining TBARS values below the unacceptable range during 20 days storage. A similar result was found by Das et al. [50], who showed that TBARS values of cooked goat meat patties with *M. oleifera* leaves extract was 47% less than control group. Sáyago-Ayerdi et al. [67] also reported similar findings using grape antioxidant dietary fibre in chicken hamburgers. In addition, this outcome agrees with those reported by other authors [9,10,68–71] who observed that the addition of natural antioxidant decrease the TBARS values of meat products.

3.4. Effect of MF on Instrumental Colour Stability during the Storage Time

The incorporation of MF extract significantly changed the colour value of treated chicken nuggets compared to control during storage time (Table 5). A significant increase in the lightness values, in treated chicken nugget (T1 and T2 groups) was observed on 0 day in comparison to control. However, the reduction in lightness value of treated chicken nuggets was noticed after 10 days of storage period which may be attributed to the dilution of meat pigment in other meat products due to the presence of non-meat ingredients.

Table 5. Effect of Moringa flower on instrumental colour of chicken nuggets during the storage time (mean values ± error standard of six samples). Control = no additive; T1 = 1.0% Moringa flower extract; T2 = 2.0% Moringa flower extract.

Treatment	Storage Time (Days)					Sig.
	0	5	10	15	20	
L * (Lightness)						
Control	29.95 ± 1.20 [ay]	26.21 ± 0.28 [b]	25.49 ± 0.23 [bcx]	23.89 ± 0.37 [cdx]	22.52 ± 0.43 [dx]	***
T1	33.80 ± 0.32 [ax]	26.21 ± 0.28 [b]	20.64 ± 0.44 [cy]	19.22 ± 0.32 [dy]	18.01 ± 0.26 [ey]	***
T2	34.78 ± 0.44 [ax]	26.21 ± 0.28 [b]	19.97 ± 0.64 [cy]	19.53 ± 0.59 [cdy]	18.51 ± 0.26 [dy]	***
Sig.	***	***	***	***	***	
a * (Redness)						
Control	13.77 ± 0.26 [ax]	12.25 ± 0.20 [bx]	6.73 ± 0.17 [cy]	5.62 ± 0.15 [cz]	5.11 ± 0.07 [dy]	***
T1	12.78 ± 0.23 [ax]	11.35 ± 0.15 [az]	7.54 ± 0.12 [bx]	6.58 ± 0.08 [by]	5.44 ± 0.16 [cy]	***
T2	11.01 ± 0.41 [ay]	10.72 ± 0.13 [by]	8.39 ± 0.08 [cx]	7.85 ± 0.10 [dx]	7.26 ± 0.16 [dx]	***
Sig.	***	***	***	***	***	
b * (Yellowness)						
Control	13.11 ± 0.26 [a]	11.81 ± 0.25 [bxy]	7.98 ± 0.07 [cx]	7.33 ± 0.13 [dx]	6.92 ± 0.17 [dx]	***
T1	13.70 ± 0.21 [a]	11.41 ± 0.13 [by]	6.30 ± 0.13 [cy]	6.44 ± 0.16 [cy]	5.68 ± 0.25 [dy]	***
T2	13.99 ± 0.52 [a]	12.30 ± 0.10 [bx]	6.45 ± 0.29 [by]	6.07 ± 0.20 [by]	6.18 ± 0.09 [by]	***
Sig.	***	***	***	***	***	

[a–e] Mean values in the same row not followed by a common letter differ significantly among storage times. [x–y] Mean values in the same column not followed by a common letter differ significantly among treatments. Sig. Significance; ns: not significant; *** $p < 0.001$.

3.5. Effect of MF on Sensory Attributes of Chicken Nuggets during the Storage Time

Although both the control and treated chicken nuggets were comparable for all the above sensory attributes up to the 5th day of storage, the control nuggets received lower sensory scores thereafter (Table 6). This finding agrees with data reported by Das et al. [49] who showed that the addition of *M. oleifera* leaves extract had no effect on the sensory attributes. In addition, Muthukumar et al. [72] noticed that the of *M. oleifera* leaves extract did not modify colour, odour, flavour or texture, and all the products were equally acceptable as evidenced by the overall acceptability scores. The sensory attributes for treated nuggets (T1 and T2 groups), even though decreased with increasing storage time, were acceptable up to the 15th day of storage. On the other hand, the sensory scores for appearance of control, T1 and T2 treatments decreased from 6.77 ± 0.15 to 5.62 ± 0.19, from 7.02 ± 0.010 to 6.44 ± 0.10 and from 6.72 ± 0.15 to 6.38 ± 0.10, respectively. There was a significant decrease ($p < 0.05$) in appearance attribute of treated nuggets from day 15 onwards; whereas, in control samples, the values were even lower on 10th day of storage time.

This fact may be due to protective effect of MF extract preventing the fading of colour of the chicken nuggets. Control group received significantly lower flavour score on day 10 and development of rancid odour was noticed on 15th day of storage time, which might be the influencing factor to reduce the flavour and acceptability scores and was, therefore, rejected by the panellists. On the other hand, MF extract containing ADF which might have acted as stabilising agent for retaining the flavour by inhibiting lipid oxidation in treated chicken nuggets. There was hardly any remarkable variation, and the texture of nuggets with MF was comparable to control ($p > 0.05$) up to 10th day. Sensory texture was not done in case of control from 15th day onwards, as rancid odour was detected. Likewise, a significant decrease ($p < 0.05$) in juiciness score was observed in both control and treated (T1 and T2) products from 10th day onwards. However, chicken nuggets containing MF extract were found to be more juicer than the control group, which could be attributed to the increased moisture retention of the product during cooking.

Table 6. Effect of Moringa flower on sensory attributes of chicken nuggets during the storage time (mean values ± error standard of six samples). Control = no additive; T1 = 1.0% Moringa flower extract; T2 = 2.0% Moringa flower extract.

Treatment	Storage Time (Days)					Sig.
	0	5	10	15	20	
Appearance						
Control	6.77 ± 0.15	6.75 ± 0.10 [xy]	6.27 ± 0.27 [aby]	6.19 ± 0.18 [b]	5.62 ± 0.19 [c]	**
T1	7.02 ± 0.10 [a]	6.97 ± 0.01 [ax]	6.94 ± 0.19 [ax]	6.69 ± 0.08 [ab]	6.44 ± 0.10 [b]	**
T2	6.72 ± 0.15	6.61 ± 0.13 [y]	6.66 ± 0.18 [y]	6.58 ± 0.12	6.38 ± 0.10	**
Sig.	ns	*	**	**	ns	
Flavour						
Control	6.84 ± 0.18 [a]	6.82 ± 0.09 [a]	6.36 ± 0.47 [b]	ND	ND	***
T1	6.97 ± 0.06 [a]	6.93 ± 0.06 [a]	6.72 ± 0.08 [ab]	6.47 ± 0.17 [bc]	6.33 ± 0.10 [c]	***
T2	6.81 ± 0.09 [a]	6.76 ± 0.09 [a]	6.70 ± 0.16 [a]	6.52 ± 0.11 [ab]	6.27 ± 0.13 [b]	***
Sig.	ns	ns	**	ns	ns	
Texture						
Control	6.72 ± 0.14	6.65 ± 0.09	6.52 ± 0.20	ND	ND	ns
T1	6.96 ± 0.10 [a]	6.79 ± 0.12 [ab]	6.50 ± 0.20 [b]	6.42 ± 0.17 [b]	6.38 ± 0.09 [b]	***
T2	6.73 ± 0.11	6.78 ± 0.13	6.82 ± 0.11	6.66 ± 0.18	6.50 ± 0.13	ns
Sig.	ns	ns	ns	ns	ns	
Juiciness						
Control	6.72 ± 0.20 [ab]	6.98 ± 0.12 [a]	6.50 ± 0.15 [b]	ND	ND	***
T1	6.80 ± 0.19	6.74 ± 0.22	6.47 ± 0.20	6.40 ± 0.12	6.30 ± 0.10	ns
T2	6.75 ± 0.10	6.75 ± 0.13	6.70 ± 0.19	6.52 ± 0.11	6.38 ± 0.10	ns
Sig.	ns	ns	ns	ns	ns	
Overall acceptability						
Control	6.80 ± 0.16 [a]	6.76 ± 0.008 [ay]	6.33 ± 0.27 [by]	ND	ND	***
T1	7.02 ± 0.07 [a]	6.91 ± 0.16 [ax]	6.73 ± 0.17 [ax]	6.66 ± 0.12 [a]	5.38 ± 0.42 [b]	***
T2	6.77 ± 0.10 [a]	6.72 ± 0.10 [ay]	6.64 ± 0.14 [abx]	6.55 ± 0.10 [ab]	6.30 ± 0.12 [b]	***
Sig.	ns	**	**	ns	ns	

ND = Not determined. [a–e] Mean values in the same row not followed by a common letter differ significantly among storage times. [x–y] Mean values in the same column not followed by a common letter differ significantly among treatments. Sig. Significance; ns: not significant; * $p < 0.05$; ** $p < 0.01$; *** $p < 0.001$.

As far as overall acceptability is concerned, the products also followed the same pattern that was observed for other sensory attributes. The control sample received a lower overall acceptability score, although nonsignificant ($p < 0.05$), on day 5 than treated samples, and was found to have rancid odour on day 15. On the other hand, overall acceptability scores obtained from T1 and T2 groups remained stable and were acceptable even on 15th day of storage. This could be due to incorporation of MF extract which might have extended the shelf-life. It is well documented by many researchers that meat products incorporated with natural antioxidants have higher flavour and overall acceptability scores during storage owing to the colour and flavour stabilising effect of them by inhibiting lipid and protein oxidation [73–79].

4. Conclusions

The results indicated that Moringa flower (MF) is a source of dietary fibre and also contains great antioxidant potential such as ferric reducing antioxidant power and radical scavenging activity. The addition of MF extract in chicken meat nuggets improved the cooking yield and dietary fibre content without affecting the acceptability of the meat product. Moreover, MF extract increased the lipid stability, odour score and shelf-life of chicken nuggets during 20 days of refrigeration storage. Therefore, MF extract could be used as a safe, natural and valuable antioxidant to the meat food industry, apart from offering its functional health promoting benefits.

Author Contributions: P.M., A.K.D., M.P., P.K.N., P.J., F.J.B. and J.M.L. conceived the idea and participated with all authors in drafting the manuscript. P.M., A.K.D., S.B., B.M., A.S. and P.K.N., elaborated the samples and analysed it. They also participate in the data interpretation, review and editing of the manuscript.

Funding: The authors would like to thank the Indian Council of Agricultural Research, New Delhi and Director, ICAR-IVRI, Izatnagar, UP, India for financial help to carry out this study. We thank Shalimar Hatcheries Limited for their help during this study.

Conflicts of Interest: The authors declare no conflicts of interest.

References

1. Lorenzo, J.M.; Pateiro, M. Influence of type of muscles on nutritional value of foal meat. *Meat Sci.* **2013**, *93*, 630–638. [CrossRef] [PubMed]

2. Larsson, S.C.; Wolk, A. Meat consumption and risk of colorectal cancer: A meta-analysis of prospective studies. *Int. J. Cancer* **2006**, *119*, 2657–2664. [CrossRef] [PubMed]

3. Dhingra, D.; Michael, M.; Rajput, H.; Patil, R.T. Dietary fibre in foods: a review. *J. Food Sci. Technol.* **2012**, *49*, 255–266. [CrossRef] [PubMed]

4. Verma, A.K.; Banerjee, R. Dietary fibre as functional ingredient in meat products: a novel approach for healthy living—A review. *J. Food Sci. Technol.* **2010**, *47*, 247–257. [CrossRef] [PubMed]

5. Bis-Souza, C.V.; Barba, F.J.; Lorenzo, J.M.; Penna, A.L.B.; Barretto, A.C.S. New strategies for the development of innovative fermented meat products: a review regarding the incorporation of probiotics and dietary fibers. *Food Rev. Int.* **2019**, *35*, 467–484. [CrossRef]

6. Verma, A.K.; Rajkumar, V.; Banerjee, R.; Biswas, S.; Das, A.K. Guava (*Psidium guajava* L.) powder as an antioxidant dietary fibre in sheep meat nuggets. *Asian-Australas. J. Anim. Sci.* **2013**, *26*, 886–895. [CrossRef] [PubMed]

7. Domínguez, R.; Gómez, M.; Fonseca, S.; Lorenzo, J.M. Effect of different cooking methods on lipid oxidation and formation of volatile compounds in foal meat. *Meat Sci.* **2014**, *97*, 223–230. [CrossRef]

8. Lorenzo, J.M.; Domínguez, R. Cooking losses, lipid oxidation and formation of volatile compounds in foal meat as affected by cooking procedure. *Flavour Fragr. J.* **2014**, *29*, 240–248. [CrossRef]

9. Domínguez, R.; Pateiro, M.; Agregán, R.; Lorenzo, J.M. Effect of the partial replacement of pork backfat by microencapsulated fish oil or mixed fish and olive oil on the quality of frankfurter type sausage. *J. Food Sci. Technol.* **2017**, *54*, 26–37. [CrossRef]

10. Domínguez, R.; Agregán, R.; Gonçalves, A.; Lorenzo, J.M. Effect of fat replacement by olive oil on the physico-chemical properties, fatty acids, cholesterol and tocopherol content of pâté. *Grasas y Aceites* **2016**, *67*, 1–9.

11. Lorenzo, J.M.; Gómez, M. Shelf life of fresh foal meat under MAP, overwrap and vacuum packaging conditions. *Meat Sci.* **2012**, *92*, 610–618. [CrossRef] [PubMed]

12. Gómez, M.; Lorenzo, J.M. Effect of packaging conditions on shelf-life of fresh foal meat. *Meat Sci.* **2012**, *91*, 513–520. [CrossRef] [PubMed]

13. Lorenzo, J.M.; Munekata, P.E.S.; Gómez, B.; Barba, F.J.; Mora, L.; Pérez-Santaescolástica, C.; Toldrá, F. Bioactive peptides as natural antioxidants in food products—A review. *Trends Food Sci. Technol.* **2018**, *79*, 136–147. [CrossRef]

14. Pateiro, M.; Bermúdez, R.; Lorenzo, J.; Franco, D. Effect of Addition of Natural Antioxidants on the Shelf-Life of "Chorizo", a Spanish Dry-Cured Sausage. *Antioxidants* **2015**, *4*, 42–67. [CrossRef] [PubMed]

15. Pateiro, M.; Lorenzo, J.; Vázquez, J.; Franco, D. Oxidation Stability of Pig Liver Pâté with Increasing Levels of Natural Antioxidants (Grape and Tea). *Antioxidants* **2015**, *4*, 102–123. [CrossRef] [PubMed]

16. Pateiro, M.; Barba, F.J.; Domínguez, R.; Sant'Ana, A.S.; Mousavi Khaneghah, A.; Gavahian, M.; Gómez, B.; Lorenzo, J.M. Essential oils as natural additives to prevent oxidation reactions in meat and meat products: A review. *Food Res. Int.* **2018**, *113*, 156–166. [CrossRef]

17. Pérez-Jiménez, J.; Serrano, J.; Tabernero, M.; Arranz, S.; Díaz-Rubio, M.E.; García-Diz, L.; Goñi, I.; Saura-Calixto, F. Bioavailability of phenolic antioxidants associated with dietary fiber: Plasma antioxidant capacity after acute and long-term intake in humans. *Plant Foods Hum. Nutr.* **2009**, *64*, 102–107. [CrossRef]

18. Das, A.K.; Rajkumar, V.; Verma, A.K. Bael Pulp Residue as a New Source of Antioxidant Dietary Fiber in Goat Meat Nuggets. *J. Food Process. Preserv.* **2015**, *39*, 1626–1635. [CrossRef]

19. Nardoia, M.; Ruiz-Capillas, C.; Herrero, A.M.; Jiménez-Colmenero, F.; Chamorro, S.; Brenes, A. Effect of added grape seed and skin on chicken thigh patties during chilled storage. *Int. J. Food Nutr. Sci.* **2017**, *4*, 67–73.

20. Falowo, A.B.; Mukumbo, F.E.; Idamokoro, E.M.; Lorenzo, J.M.; Afolayan, A.J.; Muchenje, V. Multi-functional application of *Moringa oleifera* Lam. in nutrition and animal food products: A review. *Food Res. Int.* **2018**, *106*, 317–334. [CrossRef]

21. Gopalakrishnan, L.; Doriya, K.; Kumar, D.S. *Moringa oleifera*: A review on nutritive importance and its medicinal application. *Food Sci. Hum. Wellness* **2016**, *5*, 49–56. [CrossRef]

22. Rocchetti, G.; Blasi, F.; Montesano, D.; Ghisoni, S.; Marcotullio, M.C.; Sabatini, S.; Cossignani, L.; Lucini, L. Impact of conventional/non-conventional extraction methods on the untargeted phenolic profile of *Moringa oleifera* leaves. *Food Res. Int.* **2019**, *115*, 319–327. [CrossRef] [PubMed]

23. Sankhalkar, S.; Vernekar, V. Quantitative and Qualitative analysis of Phenolic and Flavonoid content in *Moringa oleifera* Lam and *Ocimum tenuiflorum* L. *Pharmacogn. Res.* **2016**, *8*, 16–21. [CrossRef] [PubMed]

24. Singleton, V.L.; Orthofer, R.; Lamuela-Raventos, R.M. Analisys of total phenols and other oxidation sobstrates and antioxidants by means of Folin Ciocalteau reagent. *Methods Enzymol.* **1999**, *299*, 152–178.

25. Escarpa, A.; González, M. Approach to the content of total extractable phenolic compounds from different food samples by comparison of chromatographic and spectrophotometric methods. *Anal. Chim. Acta* **2001**, *427*, 119–127. [CrossRef]

26. Zeb, A. A reversed phase HPLC-DAD method for the determination of phenolic compounds in plant leaves. *Anal. Methods* **2015**, *7*, 7753–7757. [CrossRef]

27. Fargere, T.; Abdennadher, M.; Delmas, M.; Boutevin, B. Determination of peroxides and hydroperoxides with 2,2-diphenyl-1-picrylhydrazyl (DPPH). Application to ozonized ethylene vinyl acetate copolymers (EVA). *Eur. Polym. J.* **1995**, *31*, 489–497. [CrossRef]

28. Oyaizu, M. Antioxidative activities of browning reaction prepared from glucosamine. *Jpn. J. Nutr.* **1986**, *44*, 307–315. [CrossRef]

29. McCleary, B.V.; DeVries, J.W.; Rader, J.I.; Cohen, G.; Prosky, L.; Mugford, D.C.; Champ, M.; Okuma, K. Determination of Insoluble, Soluble, and Total Dietary Fiber (CODEX Definition) by Enzymatic-Gravimetric Method and Liquid Chromatography: Collaborative Study. *J. AOAC Int.* **2012**, *95*, 824–844. [CrossRef]

30. Lorenzo, J.M.; Bermúdez, R.; Domínguez, R.; Guiotto, A.; Franco, D.; Purriños, L. Physicochemical and microbial changes during the manufacturing process of dry-cured lacón salted with potassium, calcium and magnesium chloride as a partial replacement for sodium chloride. *Food Control* **2015**, *50*, 763–769. [CrossRef]

31. The Association of Official Analytical Chemists; Cunniff, P. *Official Methods of Analysis of AOAC International*, 16th ed.; The Association: Washington, DC, USA, 1995; ISBN 9780935584547.

32. Das, A.K.; Anjaneyulu, A.S.R.; Verma, A.K.; Kondaiah, N. Physicochemical, textural, sensory characteristics and storage stability of goat meat patties extended with full-fat soy paste and soy granules. *Int. J. Food Sci. Technol.* **2008**, *43*, 383–392. [CrossRef]

33. Witte, V.C.; Krause, G.F.; Bailey, M.F. A new extraction method for determining 2-thiobarbituric acid values of pork and beef during storage. *J. Food Sci.* **1970**, *35*, 582–585. [CrossRef]

34. Das, A.; Rajkumar, V.; Nanda, P.; Chauhan, P.; Pradhan, S.; Biswas, S.; Das, A.K.; Rajkumar, V.; Nanda, P.K.; Chauhan, P.; et al. Antioxidant Efficacy of Litchi (*Litchi chinensis* Sonn.) Pericarp Extract in Sheep Meat Nuggets. *Antioxidants* **2016**, *5*, 16. [CrossRef] [PubMed]

35. Arise, A.K.; Arise, R.O.; Sanusi, M.O.; Esan, O.T.; Oyeyinka, S.A. Effect of *Moringa oleifera* flower fortification on the nutritional quality and sensory properties of weaning food. *Croat. J. Food Sci. Technol.* **2014**, *6*, 65–74. [CrossRef]

36. Sánchez-Machado, D.I.; Núñez-Gastélum, J.A.; Reyes-Moreno, C.; Ramírez-Wong, B.; López-Cervantes, J. Nutritional Quality of Edible Parts of *Moringa oleifera*. *Food Anal. Methods* **2010**, *3*, 175–180. [CrossRef]

37. Blasi, F.; Urbani, E.; Cossignani, L.; Simonetti, M.S.; Chiesi, C. Seasonal variations in antioxidant compounds of *Olea europaea* leaves collected from different Italian cultivars Food Safety and Food Quality View project Structured lipid View project Seasonal variations in antioxidant compounds of *Olea europaea* leaves collected from different Italian cultivars. *Artic. J. Appl. Bot. Food Qual.* **2016**, *89*, 202–207.

38. Grigelmo-Miguel, N.; Gorinstein, S.; Martín-Belloso, O. Characterisation of peach dietary fibre concentrate as a food ingredient. *Food Chem.* **1999**, *65*, 175–181. [CrossRef]

39. Tekle, E.W.; Sahu, N.; Makesh, M. Antioxidant activitiesof moringa concanensis flowers (fresh and dried) grown in west bengal. *Int. J. Res. Chem. Environ.* **2014**, *3*, 64–70.

40. Chun, O.K.; Chung, S.J.; Song, W.O. Estimated dietary flavonoid intake and major food sources of U.S. adults. *J. Nutr.* **2007**, *137*, 1244–1252. [CrossRef]

41. Merken, H.M.; Beecher, G.R. Measurement of food flavonoids by high-performance liquid chromatography: A review. *J. Agric. Food Chem.* **2000**, *48*, 577–599. [CrossRef]

42. Chen, X.N.; Fan, J.F.; Yue, X.; Wu, X.R.; Li, L.T. Radical scavenging activity and phenolic compounds in persimmon (*Diospyros kaki* L. cv. Mopan). *J. Food Sci.* **2007**, *73*, C24–C28. [CrossRef] [PubMed]

43. Siddhuraju, P.; Abirami, A.; Nagarani, G.; Sangeethapriya, M. Antioxidant capacity and total phenolic content of aqueous acetone and ethanol extract of edible parts of *Moringa oleifera* and *Sesbania grandiflora*. *Int. J. Biol. Biomol. Agric. Food Biotechnol. Eng.* **2014**, *8*, 1090–1098.

44. Sreelatha, S.; Padma, P.R. Antioxidant activity and total phenolic content of *Moringa oleifera* leaves in two stages of maturity. *Plant Foods Hum. Nutr.* **2009**, *64*, 303–311. [CrossRef] [PubMed]

45. Alhakmani, F.; Kumar, S.; Khan, S.A. Estimation of total phenolic content, in-vitro antioxidant and anti-inflammatory activity of flowers of *Moringa oleifera*. *Asian Pac. J. Trop. Biomed.* **2013**, *3*, 623–627. [CrossRef]

46. Tekle, E.W.; Sahu, N.; Makesh, M. Antioxidative and antimicrobial activities of different solvent extracts of *Moringa oleifera*: an in vitro evaluation. *Int. J. Sci. Res. Publ.* **2015**, *5*, 255–266.

47. Devatkal, S.K.; Narsaiah, K.; Borah, A. Anti-oxidant effect of extracts of kinnow rind, pomegranate rind and seed powders in cooked goat meat patties. *Meat Sci.* **2010**, *85*, 155–159. [CrossRef] [PubMed]

48. Habib, H.; Siddiqi, R.A.; Dar, A.H.; Dar, M.A.; Gul, K.; Rashid, N.; Siddiqi, U.S. Quality characteristics of carabeef nuggets as affected by pomegranate rind powder. *J. Food Meas. Charact.* **2018**, *12*, 2164–2173. [CrossRef]

49. Das, A.K.; Rajkumar, V.; Verma, A.K.; Swarup, D. Moringa oleiferia leaves extract: a natural antioxidant for retarding lipid peroxidation in cooked goat meat patties. *Int. J. Food Sci. Technol.* **2012**, *47*, 585–591. [CrossRef]

50. Hazra, S.; Biswas, S.; Bhattacharyya, D.; Das, S.K.; Khan, A. Quality of cooked ground buffalo meat treated with the crude extracts of *Moringa oleifera* (Lam.) leaves. *J. Food Sci. Technol.* **2012**, *49*, 240–245. [CrossRef]

51. Al-Juhaimi, F.; Ghafoor, K.; Hawashin, M.D.; Alsawmahi, O.N.; Babiker, E.E. Effects of different levels of Moringa (*Moringa oleifera*) seed flour on quality attributes of beef burgers. *CYTA—J. Food* **2016**, *14*, 1–9. [CrossRef]

52. Sariçoban, C.; Özalp, B.; Yilmaz, M.T.; Özen, G.; Karakaya, M.; Akbulut, M. Characteristics of meat emulsion systems as influenced by different levels of lemon albedo. *Meat Sci.* **2008**, *80*, 599–606. [CrossRef] [PubMed]

53. Malav, O.P.; Sharma, B.D.; Kumar, R.R.; Talukder, S.; Ahmed, S.R.; Irshad, A. Quality characteristics and storage stability of functional mutton patties incorporated with red kidney bean powder. *Indian J. Small Rumin.* **2016**, *22*, 83–91. [CrossRef]

54. Choi, Y.-S.; Jeong, J.-Y.; Choi, J.-H.; Han, D.-J.; Kim, H.-Y.; Lee, M.-A.; Shim, S.-Y.; Paik, H.-D.; Kim, C.-J. Quality Characteristics of Meat Batters Containing Dietary Fiber Extracted from Rice Bran. *Korean J. Food Sci. Anim. Resour.* **2007**, *27*, 228–234. [CrossRef]

55. Ham, Y.-K.; Hwang, K.-E.; Song, D.-H.; Kim, Y.-J.; Shin, D.-J.; Kim, K.-I.; Lee, H.-J.; Kim, N.-R.; Kim, C.-J. Lotus (*Nelumbo nucifera*) rhizome as an antioxidant dietary fiber in cooked sausage: Effects on physicochemical and sensory characteristics. *Korean J. Food Sci. Anim. Resour.* **2017**, *37*, 219–227. [CrossRef] [PubMed]

56. Anderson, E.T.; Berry, B.W. Effects of inner pea fiber on fat retention and cooking yield in high fat ground beef. *Food Res. Int.* **2001**, *34*, 689–694. [CrossRef]

57. Vural, H.; Javidipour, I.; Ozbas, O.O. Effects of interesterified vegetable oils and *Sugarbeet fiber* on the quality of frankfurters. *Meat Sci.* **2004**, *67*, 65–72. [CrossRef] [PubMed]

58. Chang, H.-C.; Carpenter, J.A. Optimizing quality of Frankfurters containing oat bran and added water. *J. Food Sci.* **2006**, *62*, 194–197. [CrossRef]

59. Lorenzo, J.M.; Batlle, R.; Gómez, M. Extension of the shelf-life of foal meat with two antioxidant active packaging systems. *LWT—Food Sci. Technol.* **2014**, *59*, 181–188. [CrossRef]

60. Das, A.K.; Rajkumar, V.; Dwivedi, D.K. Antioxidant effect of curry leaf (*Murraya koenigii*) powder on quality of ground and cooked goat meat. *Int. Food Res. J.* **2011**, *18*, 563–569.

61. Dodiya, B.; Amin, B.; Kamlaben, S.; Patel, P. Antibacterial activity and phytochemical screening of different parts of *Moringa oleifera* against selected gram positive and gram negative bacteria. *J. Pharm. Chem. Biol. Sci.* **2015**, *3*, 421–425.

62. Sallam, K.I.; Ishioroshi, M.; Samejima, K. Antioxidant and antimicrobial effects of garlic in chicken sausage. *LWT—Food Sci. Technol.* **2004**, *37*, 849–855. [CrossRef] [PubMed]

63. Fernandes, R.P.P.; Trindade, M.A.; Tonin, F.G.; Lima, C.G.; Pugine, S.M.P.; Munekata, P.E.S.; Lorenzo, J.M.; de Melo, M.P. Evaluation of antioxidant capacity of 13 plant extracts by three different methods: Cluster analyses applied for selection of the natural extracts with higher antioxidant capacity to replace synthetic antioxidant in lamb burgers. *J. Food Sci. Technol.* **2016**, *53*, 451–460. [CrossRef] [PubMed]

64. Domínguez, R.; Barba, F.J.; Gómez, B.; Putnik, P.; Bursać Kovačević, D.; Pateiro, M.; Santos, E.M.; Lorenzo, J.M. Active packaging films with natural antioxidants to be used in meat industry: A review. *Food Res. Int.* **2018**, *113*, 93–101. [CrossRef] [PubMed]

65. Lorenzo, J.M.; Sineiro, J.; Amado, I.R.; Franco, D. Influence of natural extracts on the shelf life of modified atmosphere-packaged pork patties. *Meat Sci.* **2014**, *96*, 526–534. [CrossRef] [PubMed]

66. Prabakaran, M.; Kim, S.-H.; Sasireka, A.; Chandrasekaran, M.; Chung, I.-M. Polyphenol composition and antimicrobial activity of various solvent extracts from different plant parts of *Moringa oleifera*. *Food Biosci.* **2018**, *26*, 23–29. [CrossRef]

67. Sáyago-Ayerdi, S.G.; Brenes, A.; Goñi, I. Effect of grape antioxidant dietary fiber on the lipid oxidation of raw and cooked chicken hamburgers. *LWT—Food Sci. Technol.* **2009**, *42*, 971–976. [CrossRef]

68. Heck, R.T.; Fagundes, M.B.; Cichoski, A.J.; de Menezes, C.R.; Barin, J.S.; Lorenzo, J.M.; Wagner, R.; Campagnol, P.C.B. Volatile compounds and sensory profile of burgers with 50% fat replacement by microparticles of chia oil enriched with rosemary. *Meat Sci.* **2019**, *148*, 164–170. [CrossRef]

69. Munekata, P.E.S.; Paseto Fernandes, R.D.P.; de Melo, M.P.; Trindade, M.A.; Lorenzo, J.M. Influence of peanut skin extract on shelf-life of sheep patties. *Asian Pac. J. Trop. Biomed.* **2016**, *6*, 586–596. [CrossRef]

70. Agregán, R.; Franco, D.; Carballo, J.; Tomasevic, I.; Barba, F.J.; Gómez, B.; Muchenje, V.; Lorenzo, J.M. Shelf life study of healthy pork liver pâté with added seaweed extracts from Ascophyllum nodosum, Fucus vesiculosus and Bifurcaria bifurcata. *Food Res. Int.* **2018**, *112*, 400–411. [CrossRef]

71. Echegaray, N.; Gómez, B.; Barba, F.J.; Franco, D.; Estévez, M.; Carballo, J.; Marszałek, K.; Lorenzo, J.M. Chestnuts and by-products as source of natural antioxidants in meat and meat products: A review. *Trends Food Sci. Technol.* **2018**, *82*, 110–121. [CrossRef]

72. Muthukumar, M.; Naveena, B.M.; Vaithiyanathan, S.; Sen, A.R.; Sureshkumar, K. Effect of incorporation of *Moringa oleifera* leaves extract on quality of ground pork patties. *J. Food Sci. Technol.* **2014**, *5*, 3172–3180. [CrossRef] [PubMed]

73. Lorenzo, J.M.; Vargas, F.C.; Strozzi, I.; Pateiro, M.; Furtado, M.M.; Sant'Ana, A.S.; Rocchetti, G.; Barba, F.J.; Dominguez, R.; Lucini, L.; et al. Influence of pitanga leaf extracts on lipid and protein oxidation of pork burger during shelf-life. *Food Res. Int.* **2018**, *114*, 47–54. [CrossRef]

74. Pateiro, M.; Vargas, F.C.; Chincha, A.A.I.A.; Sant'Ana, A.S.; Strozzi, I.; Rocchetti, G.; Barba, F.J.; Domínguez, R.; Lucini, L.; do Amaral Sobral, P.J.; et al. Guarana seed extracts as a useful strategy to extend the shelf life of pork patties: UHPLC-ESI/QTOF phenolic profile and impact on microbial inactivation, lipid and protein oxidation and antioxidant capacity. *Food Res. Int.* **2018**, *114*, 55–63. [CrossRef] [PubMed]

75. Fernandes, R.P.P.; Trindade, M.A.; Lorenzo, J.M.; de Melo, M.P. Assessment of the stability of sheep sausages with the addition of different concentrations of Origanum vulgare extract during storage. *Meat Sci.* **2018**, *137*, 244–257. [CrossRef] [PubMed]

76. Fernandes, R.P.P.; Trindade, M.A.; Tonin, F.G.; Pugine, S.M.P.; Lima, C.G.; Lorenzo, J.M.; de Melo, M.P. Evaluation of oxidative stability of lamb burger with Origanum vulgare extract. *Food Chem.* **2017**, *233*, 101–109. [CrossRef] [PubMed]

77. Pateiro, M.; Lorenzo, J.M.; Amado, I.R.; Franco, D. Effect of addition of green tea, chestnut and grape extract on the shelf-life of pig liver pâté. *Food Chem.* **2014**, *147*, 386–394. [CrossRef] [PubMed]

78. Lorenzo, J.M.; Munekata, P.E.S.; Sant'Ana, A.S.; Carvalho, R.B.; Barba, F.J.; Toldrá, F.; Mora, L.; Trindade, M.A. Main characteristics of peanut skin and its role for the preservation of meat products. *Trends Food Sci. Technol.* **2018**, *77*, 1–10. [CrossRef]

79. Lorenzo, J.M.; González-Rodríguez, R.M.; Sánchez, M.; Amado, I.R.; Franco, D. Effects of natural (grape seed and chestnut extract) and synthetic antioxidants (buthylatedhydroxytoluene, BHT) on the physical, chemical, microbiological and sensory characteristics of dry cured sausage "chorizo". *Food Res. Int.* **2013**, *54*, 611–620. [CrossRef]

Article

Potential of a Sunflower Seed By-Product as Animal Fat Replacer in Healthier Frankfurters

Simona Grasso [1,*], Tatiana Pintado [2], Jara Pérez-Jiménez [2], Claudia Ruiz-Capillas [2] and Ana Maria Herrero [2]

[1] Institute of Food, Nutrition and Health, School of Agriculture, Policy and Development, University of Reading, Reading RG6 6AH, UK
[2] Institute of Food Science, Technology and Nutrition (ICTAN-CSIC), 28040 Madrid, Spain; tatianap@ictan.csic.es (T.P.); jara.perez@ictan.csic.es (J.P.-J.); claudia@ictan.csic.es (C.R.-C.); ana.herrero@ictan.csic.es (A.M.H.)
* Correspondence: simona.grasso@ucdconnect.ie; Tel.: +44-118-378-6576

Received: 10 March 2020; Accepted: 4 April 2020; Published: 7 April 2020

Abstract: Upcycled defatted sunflower seed flour (SUN), a by-product obtained from sunflower oil extraction, was used as an animal fat replacer to develop healthier frankfurters. For that end, animal fat was replaced (~50%) with water and 2% or 4% of SUN. Nutritional composition, technological, structural and sensorial properties were evaluated. SUN incorporation led to a significant increase in protein, minerals (magnesium, potassium, copper and manganese) and a decrease in fat content (~37% less than control with all animal fat). The incorporation of SUN in frankfurters promoted the presence of phenolic compounds. Increasing SUN addition lead to an increasingly ($p < 0.05$) darker frankfurter colour. Samples with SUN at 4% were firmer than the control according to TPA and sensory analysis results and showed the highest lipid disorder attributed to more lipid interactions in the meat matrix. SUN addition as an animal fat replacer in frankfurters is a feasible strategy to valorise sunflower oil by-products and obtain healthier frankfurters.

Keywords: by-product valorisation; sunflower meal; healthier meat product; spectroscopy analysis; polyphenol profile

---

1. Introduction

Sunflower is the fourth crop in the world for oil production after palm, soybean and rapeseed oil [1]. Sunflower meal is the main by-product of the sunflower oil production, representing up to 36% of the mass of the processed seed [2]. The protein content of sunflower seeds is about 20%, whereas the protein content of sunflower meal ranges from 30% to 50% [3]. In addition to protein, sunflower meal contains other valuable nutrients such as vitamins, minerals and polyphenols [2,4]. For this reason, although sunflower meal is mainly used as animal feed [5], it has potential for human consumption [6].

Sunflower meal has been upcycled into a versatile food-grade defatted sunflower seed powder (SUN) [7] with many potential food applications. Grasso et al. [8] showed that SUN has valuable technological properties, such as high water holding capacity. Total phenolic content and antioxidant capacity and nutritional improvements with the use of SUN in biscuits and muffins have recently been reported [8,9]. Nevertheless, the application of SUN in food from animal origin remains to be explored. The use of SUN in popular and highly consumed gel/emulsion meat products such as frankfurters is of particular interest because the SUN's reported technological properties and nutritional qualities could benefit these products formulations, obtaining healthier meat derivatives. Indeed, several meat products have been reformulated to be healthier using several strategies and ingredients to replace pork back fat [10–12]. Examples of animal fat replacements in frankfurters include the use of several plant-based ingredients (but also animal-derived ones such as collagen have been used), such as rye bran and collagen [13], chia seeds [14], pineapple dietary fibres and water [15]. Other animal fat replacements used in meat matrices are mushrooms in pork sausages [16] and grape seed oil emulsified with gelatin and alginate in meat emulsions [17]. However, to the best of our knowledge, there are no studies available on the incorporation of sunflower by-products into popular meat products such as frankfurters.

In this context, this work examined the potential of SUN application as an animal fat replacer on the reformulation of healthier frankfurters. Nutritional composition, sensory acceptability, technological properties (processing loss, colour, pH and texture) and lipid structural characteristics (using ATR–FTIR spectroscopy) were evaluated.

2. Materials and Methods

2.1. Materials

Fresh post-rigor pork meat (a mixture of *M. biceps femoris*, *M. semimembranosus*, *M. semitendinosus*, *M. gracilis* and *M. adductor*) (30 kg) (23.7% ± 0.2% protein, 3.89% ± 0.49% fat) and pork back fat (5 kg) (0.87% ± 0.1% protein, 92.99% ± 0.80% fat), both from different animals, were obtained from a local market on different days. The meat was trimmed of visible fat and connective tissue. The meat and pork backfat were minced through a grinder with a 0.6 cm plate (Van Dall SRL., Milano, Italy). Lots of approximately 1000 g for meat and 500 g for fat were vacuum packed, frozen and stored at −20 °C until used.

The SUN, donated by the company Planetarians (Palo Alto, CA, USA), was used as an animal fat replacer. This ingredient has been reported to have 38% protein, 1.8% fat and 4.6% moisture, as well as high total phenolic content [8].

Other ingredients and additives used were sodium chloride (Panreac Química, S.A., Barcelona, Spain), sodium tripolyphosphate (Manuel Riesgo, Madrid, Spain), sodium nitrite (Fulka Chemie GmbH, Buchs, Germany) and flavouring (Gewürzmüller, Julio Criado, Madrid, Spain).

2.2. Preparation of Frankfurters

Three different types of frankfurter were prepared (Table 1).

Table 1. Formulation (%) of frankfurters.

Samples *	Meat	Pork Back Fat	SUN **	Water
F/AF	61.0	15.0	0.0	21.2
F/2SUN	61.0	8.0	2.0	26.2
F/4SUN	61.0	8.0	4.0	24.2

Additives added to all the samples per 100 g of product: 2 g NaCl; 0.5 g flavouring; 0.3 g sodium tripolyphosphate and 0.012 g sodium nitrite. * Frankfurters formulated with all animal fat (F/AF), half of the animal fat and replacing the rest of fat with 2% (F/2SUN) or 4 % (F/4SUN) of defatted sunflower seed flour (SUN). ** Upcycled defatted sunflower seed flour.

One, used as a reference, was formulated with 15% of pork back fat (F/AF) to obtain a final product with about 20% of fat similar to commercial products and other frankfurters [14,15]. The other two samples were elaborated with proximately half of the pork backfat (8%). In these 2 reformulated frankfurters, to replace the rest of pork backfat (7%), SUN was used in 2 levels, 2% (F/2SUN) and 4% (F/4SUN), and the difference was made up with water, according to the formulations shown in Table 1.

Meat and fat were thawed at 3 ± 1 °C prior use. Frankfurters were elaborated according to Jiménez-Colmenero, et al. [18], using the formulation described in Table 1. Briefly, all meat was homogenized and ground for 1 min in a chilled cutter (Stephan Universal Machine UM5, Stephan Söhne GmbH and Co., Hameln, Germany). Then, pork backfat for all samples and the corresponding amount of SUN according to the formulation (Table 1) for each type of frankfurter (F/2SUN or F/4SUN), along with the rest of the ingredients (additives and water) (Table 1) were added to the cutter and mixed for 2 min. Finally, under vacuum and chilled conditions, the mixture was homogenized for

2 min. The final meat batter (<14 °C), was stuffed into 20 mm diameter Nojax cellulose casings (Viscase S.A., Bagnold Cedex, France). Samples were hand-linked and heat processed in a smokehouse (model Unimatic 1000, Micro 40 Eller, Merano, Italy). Then, frankfurters were cooled to room temperature, casings were removed, and samples were vacuum-packed in plastic bags (Cryovac® BB3050, Sealed Air S.L., Sant Boi de Llobregat, Barcelona, Spain) and stored (4 ± 1 °C) until use.

2.3. Composition Analysis and Energy Value

2.3.1. Proximate Composition

Moisture and ash contents were evaluated following the Association of Official Analytical Chemists (AOAC) [19] methods. Protein content was measured using a LECO FP-2000 Nitrogen Determinator (Leco Corporation, St Joseph, MI, USA). Fat content was evaluated according to Bligh and Dyer [20]. Fatty acid profiles were carried out in freeze-dried samples (Lyophilizer Telstar Cryodos Equipment, Tarrasa, Barcelona, Spain) by gas chromatography as reported by Pintado, et al. [21] and results were expressed as g of fatty acid per 100 g of product. All the measurements were performed in triplicate.

2.3.2. Dietary Fibre Content

Dietary fibre was determined in SUN and, due to the stability of this food constituent, dietary fibre content in the derived frankfurters was calculated according to the proportion of the ingredient. In particular, dietary fibre was measured by the indigestible fraction method [22], where the sample was subjected to several enzymatic treatments (pepsin, pancreatin, α-amylase, and amyloglucosidase) and dialysis in order to remove the digestible components of the sample and to separate soluble dietary fibre from insoluble dietary fibre. In the soluble dietary fibre, nonstarch polysaccharides were hydrolyzed with sulfuric acid and spectrophotometrically quantitated after alkalinization and reaction with dinitrosalicylic acid [23]. Insoluble dietary fibre was also subjected to hydrolysis, and nonstarch polysaccharides were measured in the supernatant by the same method, while the residue (klason lignin) was determined gravimetrically after overnight drying at 105 °C. Total dietary fibre was determined as the sum of the soluble and the insoluble dietary fibre.

2.3.3. Mineral Content

For mineral content determination, samples were prepared by acid digestion with nitric acid in a microwave digestion system (ETHOS 1, Milestone, Srl, Sorisole, Italy) [24]. The minerals were quantified on a ContrAA 700 High-Resolution Continuum Source spectrophotometer (Analytik Jena AG, Jena, Germany) equipped with a Xenon short-arc lamp (GLE, Berlin, Germany). Three determinations were carried out per sample to measure Ca, Mg, Na, K, P, Fe, Zn, Cu and Mn and results were expressed as mg/100 g product.

2.3.4. Polyphenol Content and Profile

Total extractable polyphenols in SUN and the frankfurters were extracted by means of double aqueous–organic extraction, following the method of Nardoia, et al. [25]. Extractable polyphenols were determined in the corresponding supernatants by the Folin–Ciocalteu procedure [26], using gallic acid (Sigma-Aldrich, St. Louis, MO, USA) as a standard. Results were expressed as gallic acid equivalents (mg GAE/kg of the sample).

Polyphenol profile was additionally evaluated in SUN by HPLC-ESI-QTOF MS analysis in the same extract, after concentration (6:1) with a N_2 stream, according to procedures previously described [27]. For separation, the HPLC apparatus (Agilent 1200, Agilent Technologies, Santa Clara, CA, USA) was coupled with a diode array detector (DAD) (Agilent G1315B) and a quadrupole time-of-flight (QTOF) mass analyser (Agilent G6530A) with an atmospheric pressure electrospray ionization (ESI). The column used was a 50 mm × 42.1 mm i.d., 3.5 μm, Luna C18 (Phenomenex, Torrance, CA, USA).

Gradient elution was performed with a binary system consisting of 0.1% aqueous formic acid (solvent A) and 0.1% formic acid in acetonitrile (solvent B). The following gradient was applied at a flow rate of 0.4 mL/min: 0 min, 8% B; 10 min, 323% B; 15 min, 50% B; 20 min, 50% B; 23 min, 100% B, followed by a re-equilibration step. The injection volume was 10 µL, and the column temperature was 25 °C. Data were acquired using negative ion mode with a mass range of 100–1200 Da and using a source temperature of 325 °C and a gas flow of 10 L/h. Peak identity was established by comparison with the retention times of commercial standards when available. In addition, the molecular formula proposed by the MassHunter Workstation software version 4.0 for the different signals obtained in the MS experiments were compared with previously reported phenolic compounds, especially in sunflower, and a maximum error of 10 ppm was accepted.

2.3.5. Energy Content

The energy content was calculated based on 9 kcal/g for fat; 4 kcal/g for protein and carbohydrates and 2 kcal/g for dietary fibre [28].

2.4. Composition Analysis and Energy Value

2.4.1. Processing Loss and pH

Processing loss was calculated in 10 frankfurters, as the weight loss (expressed as a percentage of initial sample weight) occurring after heat processing and chilling overnight at 2 °C. pH values were determined using an 827 Metrohm pH-meter (Metrohm AG, Zofingen, Switzerland) on homogenates of frankfurters in distilled water in a ratio of 1:10 *w/v*. For each sample, 2 homogenates were prepared on which 3 measurements were performed.

2.4.2. Colour

Surface colour determination (CIE-LAB tristimulus values, lightness, L*; redness, a* and yellowness, b*) was evaluated (Konica Minolta CM-3500d colorimeter. Konica Minolta, Madrid, Spain) on cross-sections of the samples. For each type of frankfurter, 10 determinations were carried out.

2.4.3. Texture Analysis

Textural properties were analysed by texture profile analysis (TPA) performed in a TA-XT.plus Texture Analyzer (Texture Technologies Corp., Scarsdale, NY, USA) at approximately 22 °C as described by Bourne [29]. Five cores (height = 20 mm) of each sample were axially compressed to 40% of their original height. Force–time deformation curves were obtained with a 5 kg load cell, applied at a crosshead speed of 1 mm/s. Samples were axially compressed with an aluminium cylinder probe P/32. Attributes were calculated with Texture Expert program as follows: Hardness (Hd) = peak force (N) required for first compression; cohesiveness (Ch) = ratio of active work done under the second compression curve to that done under the first compression curve (dimensionless); springiness (Sp) = distance (mm) the sample recovers after the first compression and chewiness (Cw) = Hd × Ch × Sp (N × mm).

2.5. Attenuated Total Reflectance (ATR)-FTIR Spectroscopy Analysis

The infrared spectra of each sample were recorded using a Perkin-Elmer SpectrumTM 400 spectrometer (Perkin Elmer Inc., Tres Cantos, Madrid, Spain) in mid-IR mode, equipped with an attenuated total reflectance (ATR) sampling device containing a diamond/ZnSe crystal according to Pintado, Herrero, Ruiz-Capillas, Triki, Carmona and Jimenez-Colmenero [24]. Measurements were performed at room temperature using approximately 25 mg of the samples (without any previous sample preparation), which were placed on the surface of the ATR crystal, and lightly pressed with a flat-tip plunger. For each type of frankfurter, 9 measurements were carried out. Sums of 3 spectra (24 accumulations) were performed, and these 3-sum spectra were analysed for each sample.

Spectra were acquired with the Spectrum software version 6.3.2 and spectral data were treated with the Grams/AI version 9.1 software (Thermo Electron Corporation, Waltham, MA, USA). The spectral region 3000–2800 cm^{-1} was analysed to study the lipid structural characteristics of the samples [24].

2.6. Preliminary Sensory Evaluation

The aim of the sensory test carried out was to have a preliminary evaluation of the products. In this pilot sensory study, frankfurters were assessed by a 27-member consumer panel who regularly consume similar products. Consumers were recruited at random (not targeting specific demographic groups), using mailing lists and were not compensated for taking part. Samples (2.5 cm long) from each formulation were heated for 15 s in a microwave and presented to panellists under green light to minimize colour differences between samples. Participants were instructed to evaluate the following attributes on unstructured 10 cm scales with anchored extremes: Aroma intensity (not to very), firm (not to very), juicy (not to very), powdery (not to very), flavour intensity (not to very) and overall acceptability (dislike extremely to like extremely). Each point was later converted to a numerical scale, 0 for not (aroma intensity, firm, juicy, powdery and flavour intensity) and dislike extremely (overall acceptability) and 10 for very (aroma intensity, firm, juicy, powdery and flavour intensity) and like extremely (overall acceptability). Sensory analysis was performed 2 days after frankfurters' manufacture.

2.7. Statistical Analysis

The frankfurter manufacture was repeated twice on 2 different days. One-way analysis of variance (ANOVA) was performed to evaluate differences between formulations using the SPSS program (v.22, IBM SPSS Inc., Chicago, IL, USA). Least squares differences were used for comparison of mean values between formulations and Tukey's HSD test to identify significant differences ($p < 0.05$) between formulations.

3. Results and Discussion

3.1. Composition and Energy Value

Table 2 shows the composition (proximate analysis, mineral content and total extractable polyphenols) and energy value of frankfurters.

Table 2. Proximate composition (%), mineral content (mg/100 g product), total extractable polyphenols (mg/100 g mf) and energy value (kcal/100 g product) of frankfurters.

Parameters	F/AF *	F/2SUN *	F/4SUN *
Proximate composition			
Moisture	61.54 ± 0.12 [c]	64.99 ± 0.19 [a]	64.47 ± 0.07 [b]
Protein	15.99 ± 0.04 [c]	16.96 ± 0.10 [b]	17.93 ± 0.04 [a]
Fat	18.67 ± 0.48 [a]	11.59 ± 0.26 [b]	11.75 ± 0.10 [b]
Ash	3.23 ± 0.00 [a]	2.97 ± 0.18 [a]	3.36 ± 0.12 [a]
Dietary fibre **	—	0.56	1.12
Mineral content			
Magnesium	36.32 ± 2.73 [c]	60.72 ± 8.17 [b]	85.50 ± 3.20 [a]
Potassium	302.41 ± 10.44 [c]	383.34 ± 2.98 [b]	415.16 ± 20.40 [a]
Iron	1.18 ± 0.09 [b]	1.57 ± 0.16 [ab]	2.02 ± 0.18 [a]
Zinc	2.02 ± 0.04 [b]	2.28 ± 0.15 [b]	2.95 ± 0.16 [a]
Copper	0.06 ± 0.01 [c]	0.14 ± 0.01 [b]	0.21 ± 0.01 [a]
Manganese	0.06 ± 0.00 [c]	0.18 ± 0.00 [b]	0.29 ± 0.01 [a]
Polyphenols			
Total extractable polyphenols (mg/100 g mf)	51 ± 9 [c]	76 ± 11 [b]	93 ± 12 [a]
Energy value	231.99	173.21	179.59

* For samples denominations, see Table 1. ** Dietary fibre content was estimated according to the content analysed in SUN, so no statistical analysis was performed. Data are expressed as means ± SD ($n = 6$). Different letters in the same row indicate significant differences ($p < 0.05$).

As expected, samples formulated with all pork backfat (F/AF) had the lowest ($p < 0.05$) moisture content, because they were formulated with the lowest water content (Table 1). It was observed that with the SUN addition, protein content significantly increased (Table 2) with increasing SUN addition, due to the protein content of this ingredient [8]. As a consequence of the use of SUN extract as an animal fat replacer, two fat levels were observed, ~19% in samples elaborated with all animal fat (F/AF) and nearer to 12% in frankfurters reformulated with SUN (F/2SUN and F/4SUN) (Table 2). This fat reduction meant a decrease in saturated fatty acids, as can be seen in Figure 1. The unsaturated fatty acids content also was reduced as a consequence of the use of SUN extract as an animal fat replacer in F/2SUN and F/4SUN samples (Figure 1). Overall the SUN addition led to a quantitative fatty acid modification, rather than to a qualitative change in fatty acids.

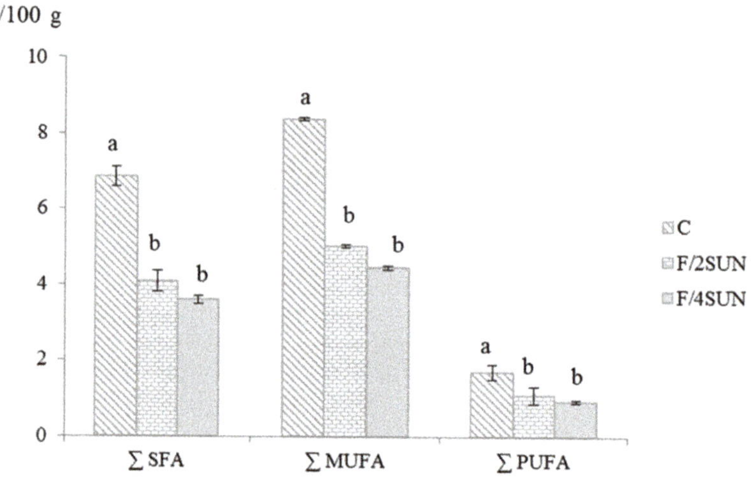

Figure 1. Σ SFA, Σ MUFA and Σ PUFA of frankfurters (g/100 g of frankfurter). For sample denomination, see Table 1. Different letters in the same fatty acid group indicate significant differences ($p < 0.05$).

Regarding dietary fibre, SUN showed a relevant content of 28% in the freeze-dried sample, of which 87% corresponded to insoluble dietary fibre. Since dietary fibre intakes are still far from official recommendations [30], this shows the potential of SUN as a functional ingredient for different food applications. Due to the amount of SUN incorporated into the frankfurters, the final dietary fibre content in F/2SUN was 0.56%, while in F2/4SUN was 1.12%, representing low amounts, although higher than in F/AF, where dietary fibre was absent.

Although absolute values were different for ash content among samples, there was no statistical difference to report (Table 2). Ash content is a good indicator of total mineral content and according to the literature [31], sunflower oil cake on a dry basis contains 0.48% calcium, 0.84% phosphorus, 0.44% magnesium and 3.49% potassium. Not all minerals were analysed in the samples, but the content of some minerals was improved as a consequence of SUN addition in F/2SUN and F/4SUN samples (Table 2). Both F/2SUN and F/4SUN showed higher ($p < 0.05$) magnesium, potassium, copper and manganese concentration than reference samples (F/AF). Regarding zinc, F/2SUN and F/AF showed similar ($p > 0.05$) levels, while F/4SUN had significantly higher amounts. Iron levels increased as a consequence of SUN addition, but the increase was significant only in F/4SUN compared to the control; this was probably due to the fact that most of the iron came from the meat materials and, therefore, only the higher SUN inclusion made a significant difference to the iron content.

Energy value in the control samples (F/AF) was ~232 kcal/100 g, while reformulated samples with SUN, F/2SUN and F/4SUN, showed an energy value of proximately 173 and 180 kcal/100 g, respectively

(Table 2). In that sense, the reformulation strategy carried out allowed to obtain products with ~25% less of energy value than control one. This difference in energy value was mainly due to the lower fat content (but also, the higher fibre, water and protein contents) of samples with SUN compared to F/AF.

3.2. Polyphenol Content and Profile

Total extractable polyphenol content was determined in the analysed samples (Table 2). A value was obtained for the F/AF samples, despite the absence of SUN, since it is known that in matrixes of animal origin several interfering substances may provide a response to the Folin assay, as observed, for instance, in minced fish [32]. Nevertheless, the most relevant fact is that there was a significant increase in polyphenol content when adding SUN to the frankfurters, even at relatively low concentrations such as 2% and 4%. Whether this presence of phenolic compounds would delay the oxidation of frankfurters remains to be elucidated.

Besides, a total of 24 phenolic compounds were identified in SUN (Table 3): 18 phenolic acids, 5 flavonoids and 1 hydroxycoumarin. Phenolic acids, and particularly quinic derivatives of caffeic or ferulic acids, have been previously reported as the main phenolic compounds in sunflower kernel and shell [33] and, specifically, in sunflower meal [34,35]. These acids were detected in this study, together with some additional phenolic acids (hydroxybenzoic acid, vanillic acid, sinapic acid, a quinolactone from caffeic acid, 1,2-disinapoylgentiobiose and the hydroxyphenylpropionic acid 3,4-dihydroxyphenyl-2-oxypropanoic acid), which, as far as we know, had not been reported in sunflower products before. Regarding the other phenolic compounds, some of them are quite ubiquitous in plant foods, such as (+)-gallocatechin while, for instance, 5-hydroxy-4,4′,6-trimethoxyaureone has not been reported in sunflower seeds, but it has in sunflower flowers as well as in different oils; for these reasons, their presence in SUN was plausible.

Table 3. Phenolic compounds identified in upcycled defatted sunflower seed flour by HPLC-ESI-QTOF MS analysis.

ID	Class	Subclass	Proposed Compound	Experimental Mass (M-H)$^-$	Calculated Mass (M-H)$^-$	Error (ppm)	Molecular Formula
1	Phenolic acids	Hydroxybenzoic acids	Hydroxybenzoic acid	137.023	137.0244	10.27	$C_7H_6O_3$
2	Phenolic acids	Hydroxybenzoic acids	Vanillic acid	167.0344	167.0350	3.47	$C_8H_8O_4$
3	Phenolic acids	Hydroxycinnamic acids	Caffeic Acid	179.0354	179.0350	2.32	$C_9H_8O_4$
4	Phenolic acids	Hydroxycinnamic acids	1-O-Caffeoylquinic Acid	353.0864	353.0878	3.27	$C_{16}H_{18}O_9$
5	Phenolic acids	Hydroxycinnamic acids	3-O-Caffeoylquinic acid	353.0915	353.0878	10.43	$C_{16}H_{18}O_9$
6	Phenolic acids	Hydroxycinnamic acids	4-O-Caffeoylquinic acid	353.0849	353.0878	8.21	$C_{16}H_{18}O_9$
7	Phenolic acids	Hydroxycinnamic acids	5-O-Caffeoylquinic acid	353.0899	353.0878	5.91	$C_{16}H_{18}O_9$
8	Phenolic acids	Hydroxycinnamic acids	3,4-di-o-caffeoylquinic acid	515.1230	515.1195	6.78	$C_{25}H_{24}O_{12}$
9	Phenolic acids	Hydroxycinnamic acids	3,5-di-o-caffeoylquinic acid	515.1200	515.1195	0.97	$C_{25}H_{24}O_{12}$
10	Phenolic acids	Hydroxycinnamic acids	4,5-di-o-caffeoylquinic acid	515.1211	515.1195	3.10	$C_{25}H_{24}O_{12}$
11	Phenolic acids	Hydroxycinnamic acids	Caffeoyl-1,5-quinolactone	335.0779	335.0772	1.96	$C_{16}H_{16}O_8$
12	Phenolic acids	Hydroxycinnamic acids	5-O-*p*-Coumaroylquinic acid	337.0949	337.0929	5.94	$C_{16}H_{18}O_8$
13	Phenolic acids	Hydroxycinnamic acids	Feruloylquinic acid	367.1060	367.1035	6.91	$C_{17}H_{20}O_9$
14	Phenolic acids	Hydroxycinnamic acids	Feruloylquinic acid	367.1060	367.1035	6.91	$C_{17}H_{20}O_9$
15	Phenolic acids	Hydroxycinnamic acids	Sinapic acid	223.0605	223.0612	3.11	$C_{11}H_{12}O_5$
16	Phenolic acids	Hydroxycinnamic acids	Hydroxycaffeic acid	195.0315	195.0299	8.18	$C_9H_8O_5$
17	Phenolic acids	Hydroxycinnamic acids	1,2-Disinapoylgentiobiose	753.2272	753.2248	3.24	$C_{34}H_{42}O_{19}$
18	Phenolic acids	Hydroxyphenylpropanoic	3,4-dihydroxyphenyl-2-oxypropanoic acid	179.0354	179.0350	2.32	$C_9H_8O_4$
19	Flavonoids	Aurone flavonoids	5-Hydroxy-4,4′,6-trimethoxyaurone	327.1097	327.1085	0.13	$C_{18}H_{16}O_6$
20	Flavonoids	Dihydroflavonols	Dihydroquercetin 3-O-rhamnoside	449.1089	449.1089	0.08	$C_{21}H_{22}O_{11}$
21	Flavonoids	Flavanols	(+)-Gallocatechin	305.0695	305.0667	5.96	$C_{15}H_{14}O_7$
22	Flavonoids	Flavonoid glycosides	Didymin	593.1889	593.1876	2.22	$C_{28}H_{34}O_{14}$
23	Flavonoids	Flavonols	Isorhamnetin hexoside	477.1051	477.1038	2.62	$C_{22}H_{22}O_{12}$
24	Other polyphenols	Hydroxycoumarins	Esculin	339.0714	339.0722	2.22	$C_{15}H_{16}O_9$

3.3. Overall Nutritional Value: Nutrition and Health Claims

All samples were a "high in protein" and could be labelled with the corresponding nutritional and health claims according to the Regulation (EC) n° 1924/2006 and Regulation (EU) n° 432/2012 [36,37]. Moreover, as a consequence of the use of SUN as an animal fat replacer, other nutrition and health claims can be made. Taking into account the fat content, the claim "reduced fat content" in F/2SUN and F/4SUN can be made [36]. The increased mineral content due to the presence of SUN, makes it possible for the "source of magnesium" claim to be made on F/2SUN and F/4SUN. In addition, several health claims could be attributed to F/2SUN and F/4SUN according to Regulation (EU) n° 432/2012, due to the magnesium presence. Moreover, F/4SUN could be labelled as "source of copper" and the health claims corresponding to its presence. Regarding zinc and potassium, although their presence was generally higher in F/2SUN and F/4SUN (Table 2), all samples could be labelled as "source of zinc" and "source of potassium" and related health claims [36,37]. These results were similar to other studies where pork back fat in frankfurters was replaced with other high protein ingredients such as hydrolysed collagen [38], and chia seeds [14].

3.4. Processing Loss, pH and Colour

The processing loss and pH results are shown in Table 4.

Table 4. Processing loss, pH values, colour and texture profile analysis (TPA) of frankfurters.

Parameters	F/AF *	F/2SUN *	F/4SUN *
Processing loss (%)	15.56 ± 1.34 [a]	16.33 ± 1.10 [a]	15.46 ± 1.72 [a]
pH	6.39 ± 0.02 [a]	6.37 ± 0.05 [b]	6.34 ± 0.03 [c]
Colour parameters			
L*	76.37 ± 0.50 [a]	62.33 ± 1.09 [b]	57.14 ± 0.88 [c]
a*	6.37 ± 0.27 [a]	3.24 ± 0.12 [b]	2.88 ± 0.14 [c]
b*	11.32 ± 0.31 [a]	10.21 ± 0.21 [b]	10.29 ± 0.27 [b]
TPA parameters			
Hardness (N)	17.52 ± 0.92 [b]	18.27 ± 1.49 [b]	21.67 ± 2.32 [a]
Cohesiveness	0.64 ± 0.01 [b]	0.67 ± 0.01 [a]	0.67 ± 0.01 [a]
Springiness (mm)	6.11 ± 0.08 [b]	6.36 ± 0.08 [a]	6.41 ± 0.08 [a]
Chewiness (mm × N)	68.03 ± 3.19 [c]	77.99 ± 6.06 [b]	92.35 ± 9.82 [a]

* For samples denominations, see Table 1. Data are expressed as means ± SD ($n = 10$ for processing loss, $n = 3$ for pH values, $n = 20$ for colour parameters and $n = 8$ for texture parameters). Different letters in the same row indicate significant differences ($p < 0.05$).

There was no significant difference in the processing loss across the three recipes. These results were low and similar to other frankfurters' studies, which reported ranges of processing loss between 10% and 20% [39–41]. The processing loss obtained (Table 4) indicated good stability in terms of fat and water-binding properties of the meat matrix. The pH decreased with increasing SUN content. The SUN addition influenced the colour, with L*, a* and b* values decreasing significantly with increased SUN content, indicating that the frankfurters became darker, less red and less yellow. It is important that future SUN applications are aimed at minimising colour differences in the food matrix to avoid possible consumer's rejection, as liking of foods with adequate appearance could favour healthier product consumption by consumers [42]. During the process of new product development, the colour of foods with SUN could be altered to obtain no noticeable colour difference for consumers. Colour alterations have been reported in other studies where pork back fat in frankfurters was replaced by chia seeds [14] and pineapple fibre [15] as well as in other studies using SUN in baked goods [8,9].

3.5. Texture Profile Analysis

The results from the texture profile analysis are shown in Table 4. Results showed that hardness was similar ($p > 0.05$) in frankfurters containing 2% of SUN (F/2SUN) but significantly higher in those with 4% (F/4SUN) as compared with the control (F/AF) (Table 4). This is probably due to the water holding capacity of the SUN [8]. The SUN might have retained water, swollen and consequently increased firmness, similarly to the behaviour reported on frankfurters with collagen [13,38,43]. Cohesiveness, springiness and chewiness were higher ($p < 0.05$) in samples with SUN added (F/2SUN and F/4SUN), regardless of the amount of this ingredient used as animal fat replacer (Table 4). This textural behaviour could be related with the differences in composition (mainly protein and fibre content) in each type of frankfurter due to of addition of SUN as an animal fat replacer, although the muscle protein level was kept constant in the formulations.

3.6. Attenuated Total Reflectance (ATR)-FTIR Spectroscopy Analysis

The acyl chain region comprised between 2950–2830 cm^{-1} of the ATR–FTIR spectrum of the different frankfurter are shown in Figure 2.

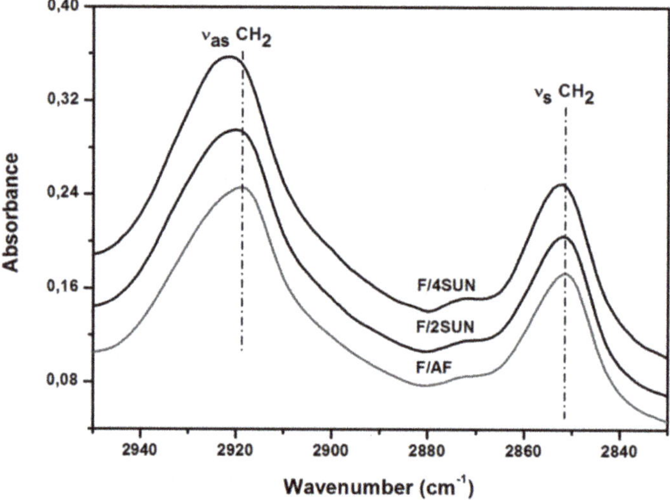

Figure 2. ATR–FTIR spectra in the 2950–2830 cm^{-1} of frankfurters. For sample denomination, see Table 1.

This spectral region was dominated by two strong bands resulting, respectively, from the asymmetric ($_{as}$CH$_2$) and the symmetric ($_s$CH$_2$) stretching vibrations of the acyl CH$_2$ groups [44]. Partial replacement of animal fat by SUN produces a frequency upshift in $_{as}$CH$_2$ and $_s$CH$_2$ from 2919 to 2921 cm^{-1} and from 2851 to 2852 cm^{-1}, respectively, in going to from F/AF to F/4SUN (Figure 2). This frequency upshift was generally attributed to the diminution of the conformational order of the lipid acyl chains and to the increase of their dynamics, which implied greater inter- and intramolecular lipid disorder [45,46]. Therefore, it is possible to assume that frankfurters elaborated with SUN showed greater inter- and intramolecular lipid disorder than control elaborated with all animal fat, being these phenomena more relevant in F/4SUN samples. The increase of lipid disorder in reformulated frankfurters as a function of SUN content could be attributed to more lipid interactions in the meat matrix in these samples (F/2SUN and F/4SUN), mainly protein-lipid interactions [47]. The lipid chain disorder or increased lipid interactions observed in F/2SUN and F/4SUN could be related with their specific textural behaviour (Table 3), greater hardness, springiness and chewiness than control (F/AF).

Accordingly, previous studies showed that the direct addition of vegetable proteins, such as soy protein isolate, in the reformulation process of heated meat batters produced protein secondary structural changes accompanied by textural properties modifications such as an increase in hardness and chewiness [48]. In addition, it has been described that the direct addition of mushroom powder, with similar composition as SUN in terms of protein, dietary fibre and bioactive compound content, in the reformulation of processed meats impacts on their rheological and structural characteristics. Similar to results found in the present work, it has been indicated that this reformulation process increased the conformational disorder of the lipid acyl chains due to modifications of lipid-protein interactions [49].

3.7. Preliminary Sensory Analysis

The results of the pilot sensory study are shown in Table 5.

Table 5. Pilot sensory analysis of frankfurters.

Parameters	F/AF *	F/2SUN *	F/4SUN *
Aroma intensity	3.32 ± 2.34 [a]	4.19 ± 2.53 [a]	4.76 ± 2.63 [a]
Firmness	6.10 ± 1.76 [b]	6.89 ± 1.51 [ab]	7.82 ± 2.00 [a]
Juiciness	5.42 ± 2.29 [a]	4.70 ± 2.43 [ab]	3.31 ± 2.34 [b]
Powdery	1.67 ± 1.51 [a]	2.53 ± 2.54 [a]	3.00 ± 2.39 [a]
Flavour intensity	5.45 ± 2.21 [a]	5.56 ± 2.15 [a]	6.54 ± 1.98 [a]
Overall acceptability	6.07 ± 2.08 [a]	5.85 ± 2.38 [a]	4.87 ± 2.60 [a]

* For samples denominations, see Table 1. Data are expressed as means ± SD ($n = 27$). Different letters in the same row indicate significant differences ($p < 0.05$).

The panel detected a significant difference only in the textural sensory parameter's firmness and juiciness. F/4SUN samples were considered firmer than F/AF, which agrees with findings from the texture profile analysis. F/4SUN samples were also considered significantly less juicy than F/AF. This might be related to another attribute, powdery, which tended to increase in samples with SUN, although not significantly. Samples with SUN tended to have a higher aroma and flavour intensity and lower overall acceptability than control, but these differences were not significant. The reduction of pork back fat in frankfurters has been associated with lower acceptability scores in several sensory attributes [14,16]. Since a limited number of consumers was used to in the present study, sensory tests run on a bigger sample size would need to be repeated in the future in order to ensure enough statistical power.

4. Conclusions

The incorporation of SUN as an animal fat replacer in frankfurters offers interesting possibilities for the valorisation of sunflower oil by-products. SUN addition led to several nutritional improvements, including the lipid content (qualitative and quantitative), an increase in protein and minerals. SUN inclusion in frankfurters also promoted the presence of phenolic compounds, mostly phenolic acids. Frankfurters with SUN could benefit from several on-pack nutrition and health claims. Although the reformulation process using SUN as animal fat replacer resulted in some changes in the technological and structural characteristics of the products, the products were found acceptable from a technological point of view. Further sensory tests with a larger consumers' sample will be needed to build onto the preliminary results from the pilot sensory study here reported. In the future, SUN could also be used as an animal fat alternative in plant-based meat substitutes or cell-cultured meat products as part of an effort to replace animal products and have a positive impact on animal welfare and the environment.

Author Contributions: Conceptualization, S.G., T.P., C.R.-C., and A.M.H.; data curation, S.G., T.P., J.P.-J., and A.M.H.; formal analysis, S.G., T.P., J.P.-J., and A.M.H.; funding acquisition, C.R.-C., and A.M.H.; investigation, S.G., T.P., C.R.-C., and A.M.H.; methodology, S.G., T.P., J.P.-J., C.R.-C., and A.M.H.; project administration, C.R.-C., and A.M.H.; resources, C.R.-C., and A.M.H.; supervision, C.R.-C; validation, J.P.-J; visualization, A.M.H.; writing—Original draft, S.G., T.P., J.P.-J., C.R.-C., and A.M.H.; writing—Review and editing, S.G., T.P., J.P.-J., C.R.-C., and A.M.H. All authors have read and agreed to the published version of the manuscript.

Funding: This research was funded by the CSIC Intramural project (grant number 201470E073) and CYTED (grant number reference 119RT0568; HealthyMeat network). Anastasia Palatzidi was the recipient of an Erasmus+ fellowship and of an EIT Food RIS Fellowship 2019—funded by the European Institute of Innovation and Technology, a body of the European Union. Simona Grasso was a recipient of a travel bursary from the Institute of Food, Nutrition and Health at the University of Reading.

Acknowledgments: Andrea Sánchez-Monedero and Anastasia Palatzidi are acknowledged for technical assistance during polyphenols and dietary fibre analyses, respectively. We are grateful to the ICTAN Analysis Services Unit for providing facilities for chromatography analysis and in particular to Inmaculada Álvarez-Acero for technical assistance. The authors are grateful to the company Planetarians for donating the defatted sunflower seed flour.

Conflicts of Interest: The authors declare no conflict of interest. The funders had no role in the design of the study; in the collection, analyses, or interpretation of data; in the writing of the manuscript, or in the decision to publish the results.

References

1. FAO. *FAOSTAT Statistics Database—Agriculture*; FAO: Rome, Italy, 2018; Available online: http://www.fao.org/faostat/en/#data/QC (accessed on 6 April 2020).

2. Yegorov, B.; Turpurova, T.; Sharabaeva, E.; Bondar, Y. Prospects of using by-products of sunflower oil production in compound feed industry. *J. Food Sci. Technol.* **2019**, *13*, 106–113. [CrossRef]

3. Dorrell, D.G.; Vick, B.A. Properties and processing of oilseed sunflower. *Sunflower Technol. Prod.* **1997**, *35*, 709–745.

4. Wanjari, N.; Waghmare, J. Phenolic and antioxidant potential of sunflower meal. *Adv. Appl. Sci. Res.* **2015**, *6*, 221–229.

5. Anal, A.K. *Food Processing By-Products and Their Utilization*; John Wiley & Sons: Hoboken, NJ, USA, 2017.

6. Salgado, P.R.; Molina Ortiz, S.E.; Petruccelli, S.; Mauri, A.N. Functional food ingredients based on sunflower protein concentrates naturally enriched with antioxidant phenolic compounds. *J. Am. Oil Chem. Soc.* **2012**, *89*, 825–836. [CrossRef]

7. Manchuliantsau, A.; Tkacheva, A. Upcycling Solid Food Wastes and By-Products into Human Consumption Products. U.S. Patent 16,370,896, 25 July 2019.

8. Grasso, S.; Omoarukhe, E.; Wen, X.; Papoutsis, K.; Methven, L. The use of upcycled defatted sunflower seed flour as a functional ingredient in biscuits. *Foods* **2019**, *8*, 305. [CrossRef] [PubMed]

9. Grasso, S.; Liu, S.; Methven, L. Quality of muffins enriched with upcycled defatted sunflower seed flour. *LWT* **2020**, *119*, 108893. [CrossRef]

10. Jimenez-Colmenero, F.; Salcedo-Sandoval, L.; Bou, R.; Cofrades, S.; Herrero, A.M.; Ruiz-Capillas, C. Novel applications of oil-structuring methods as a strategy to improve the fat content of meat products. *Trends Food Sci. Technol.* **2015**, *44*, 177–188. [CrossRef]

11. Weiss, J.; Gibis, M.; Schuh, V.; Salminen, H. Advances in ingredient and processing systems for meat and meat products. *Meat Sci.* **2010**, *86*, 196–213. [CrossRef]

12. Jiménez-Colmenero, F. Healthier lipid formulation approaches in meat-based functional foods. Technological options for replacement of meat fats by non-meat fats. *Trends Food Sci. Technol.* **2007**, *18*, 567–578.

13. Hjelm, L.; Mielby, L.A.; Gregersen, S.; Eggers, N.; Bertram, H.C. Partial substitution of fat with rye bran fibre in Frankfurter sausages—Bridging technological and sensory attributes through inclusion of collagenous protein. *LWT* **2019**, *101*, 607–617. [CrossRef]

14. Pintado, T.; Herrero, A.M.; Jiménez-Colmenero, F.; Ruiz-Capillas, C. Strategies for incorporation of chia (Salvia hispanica L.) in frankfurters as a health-promoting ingredient. *Meat Sci.* **2016**, *114*, 75–84. [CrossRef]

15. Henning, S.S.C.; Tshalibe, P.; Hoffman, L.C. Physico-chemical properties of reduced-fat beef species sausage with pork back fat replaced by pineapple dietary fibres and water. *LWT* **2016**, *74*, 92–98. [CrossRef]

16. Wang, L.; Li, C.; Ren, L.; Guo, H.; Li, Y. Production of pork sausages using pleaurotus eryngii with different treatments as replacements for pork back fat. *J. Food Sci.* **2019**, *84*, 3091–3098. [CrossRef]

17. Kim, T.-K.; Yong, H.-I.; Jung, S.; Kim, Y.-B.; Choi, Y.-S. Effects of replacing pork fat with grape seed oil and gelatine/alginate for meat emulsions. *Meat Sci.* **2020**, *163*, 108079. [CrossRef]

18. Jiménez-Colmenero, F.; Herrero, A.; Pintado, T.; Solas, M.T.; Ruiz-Capillas, C. Influence of emulsified olive oil stabilizing system used for pork backfat replacement in frankfurters. *Food Res. Int.* **2010**, *43*, 2068–2076. [CrossRef]

19. AOAC. *Official Methods of Analysis*, 18th ed.; Association of Official Analytical Chemists: Washington, DC, USA, 2005.

20. Bligh, E.G.; Dyer, W.J. A rapid method of total lipid extraction and purification. *Can. J. Biochem. Physiol.* **1959**, *37*, 911–917. [CrossRef] [PubMed]

21. Pintado, T.; Herrero, A.M.; Jiménez-Colmenero, F.; Cavalheiro, C.P.; Ruiz-Capillas, C. Chia and oat emulsion gels as new animal fat replacers and healthy bioactive sources in fresh sausage formulation. *Meat Sci.* **2018**, *135*, 6–13. [CrossRef] [PubMed]

22. Goñi, I.; Díaz-Rubio, M.E.; Pérez-Jiménez, J.; Saura-Calixto, F. Towards an updated methodology for measurement of dietary fiber, including associated polyphenols, in food and beverages. *Food Res. Int.* **2009**, *42*, 840–846. [CrossRef]

23. Englyst, H.N.; Cummings, J. Improved method for measurement of dietary fiber as non-starch polysaccharides in plant foods. *J. Assoc. Off. Anal. Chem.* **1988**, *71*, 808–814. [CrossRef] [PubMed]

24. Pintado, T.; Herrero, A.; Ruiz-Capillas, C.; Triki, M.; Carmona, P.; Jimenez-Colmenero, F. Effects of emulsion gels containing bioactive compounds on sensorial, technological, and structural properties of frankfurters. *Food Sci. Technol. Int.* **2016**, *22*, 132–145. [CrossRef]

25. Nardoia, M.; Ruiz-Capillas, C.; Casamassima, D.; Herrero, A.M.; Pintado, T.; Jiménez-Colmenero, F.; Chamorro, S.; Brenes, A. Effect of polyphenols dietary grape by-products on chicken patties. *Eur. Food Res. Technol.* **2018**, *244*, 367–377. [CrossRef]

26. Singleton, V.L.; Orthofer, R.; Lamuela-Raventos, R.M. Section III. Polyphenols and flavonoids-14-analysis of total phenols and other oxidation substrates and antioxidants by means of folin-ciocalteu reagent. *Methods Enzymol.* **1998**, *1999*, 152–177.

27. Pérez-Ramírez, I.F.; Reynoso-Camacho, R.; Saura-Calixto, F.; Pérez-Jiménez, J. Comprehensive characterization of extractable and nonextractable phenolic compounds by high-performance liquid chromatography–electrospray ionization–quadrupole time-of-flight of a grape/pomegranate pomace dietary supplement. *J. Agric. Food Chem.* **2018**, *66*, 661–673. [CrossRef] [PubMed]

28. EU Regulation. No 1169/2011 of the European Parliament and of the Council of 25 October 2011 on the provision of food information to consumers. *Eur. Comm. Off. J. Eur. Union* **2011**, *20*, 168–213.

29. Bourne, M.C. Texture profile analysis. *Food Technol.* **1978**, *32*, 62–66.

30. GBD 2017 Diet Collaborators. Health effects of dietary risks in 195 countries, 1990–2017: A systematic analysis for the global burden of disease study 2017. *Lancet* **2019**, *393*, 1958–1972. [CrossRef]

31. Ratcliff, R.K. Nutritional Value of Sunflower Meal for Ruminants. Ph.D. Thesis, Texas Tech University, Lubbock, TX, USA, 1977.

32. Solari-Godiño, A.; Pérez-Jiménez, J.; Saura-Calixto, F.; Borderías, A.J.; Moreno, H. Anchovy mince (Engraulis ringens) enriched with polyphenol-rich grape pomace dietary fibre: In vitro polyphenols bioaccessibility, antioxidant and physico-chemical properties. *Food Res. Int.* **2017**, *102*, 639–646. [CrossRef]

33. Weisz, G.M.; Kammerer, D.R.; Carle, R. Identification and quantification of phenolic compounds from sunflower (Helianthus annuus L.) kernels and shells by HPLC-DAD/ESI-MSn. *Food Chem.* **2009**, *115*, 758–765. [CrossRef]

34. Laguna, O.; Barakat, A.; Alhamada, H.; Durand, E.; Baréa, B.; Fine, F.; Villeneuve, P.; Citeau, M.; Dauguet, S.; Lecomte, J. Production of proteins and phenolic compounds enriched fractions from rapeseed and sunflower meals by dry fractionation processes. *Ind. Crop. Prod.* **2018**, *118*, 160–172. [CrossRef]

35. Slabi, S.A.; Mathé, C.; Framboisier, X.; Defaix, C.; Mesieres, O.; Galet, O.; Kapel, R. A new SE-HPLC method for simultaneous quantification of proteins and main phenolic compounds from sunflower meal aqueous extracts. *Anal. Bioanal. Chem.* **2019**, *411*, 2089–2099. [CrossRef]

36. European Commission. Regulation (EC) No 1924/2006 of the European Parliament and of the Council of 20 December 2006 on nutrition and health claims made on foods. *Eur. Comm. Off. J. Eur. Union* **2006**, *404*, 3–18.

37. European Commission. Regulation (EU) No 432/2012 of the European Parliament and of the Council of 16 may 2012 establishing a list of permited health claims made on foods other than those referring to the reduction of disease risk and to children's development and health. *Eur. Comm. Off. J. Eur. Union* **2012**, *136*, 1–40.

38. Sousa, S.C.; Fragoso, S.P.; Penna, C.R.A.; Arcanjo, N.M.O.; Silva, F.A.P.; Ferreira, V.C.S.; Barreto, M.D.S.; Araújo, Í.B.S. Quality parameters of frankfurter-type sausages with partial replacement of fat by hydrolyzed collagen. *LWT-Food Sci. Technol.* **2017**, *76*, 320–325. [CrossRef]

39. Herrero, A.M.; Ruiz-Capillas, C.; Pintado, T.; Carmona, P.; Jimenez-Colmenero, F. Infrared spectroscopy used to determine effects of chia and olive oil incorporation strategies on lipid structure of reduced-fat frankfurters. *Food Chem.* **2017**, *221*, 1333–1339. [CrossRef] [PubMed]

40. Pintado, T.; Ruiz-Capillas, C.; Jiménez-Colmenero, F.; Carmona, P.; Herrero, A.M. Oil-in-water emulsion gels stabilized with chia (Salvia hispanica L.) and cold gelling agents: Technological and infrared spectroscopic characterization. *Food Chem.* **2015**, *185*, 470–478. [CrossRef] [PubMed]

41. López-López, I.; Cofrades, S.; Jiménez-Colmenero, F. Low-fat frankfurters enriched with n−3 PUFA and edible seaweed: Effects of olive oil and chilled storage on physicochemical, sensory and microbial characteristics. *Meat Sci.* **2009**, *83*, 148–154. [CrossRef] [PubMed]

42. Deliza, R.; Saldivar, S.S.; Germani, R.; Benassi, V.; Cabral, L. The effects of colored textured soybean protein (TSP) on sensory and physical attributes of ground beef patties. *J. Sens. Stud.* **2002**, *17*, 121–132. [CrossRef]

43. Pereira, A.G.T.; Ramos, E.M.; Teixeira, J.T.; Cardoso, G.P.; Ramos, A.D.L.S.; Fontes, P.R. Effects of the addition of mechanically deboned poultry meat and collagen fibers on quality characteristics of frankfurter-type sausages. *Meat Sci.* **2011**, *89*, 519–525. [CrossRef]

44. Guillen, M.D.; Cabo, N. Infrared spectroscopy in the study of edible oils and fats. *J. Sci. Food Agric.* **1997**, *75*, 1–11. [CrossRef]

45. Fraile, M.; Patrón-Gallardo, B.; López-Rodrıguez, G.; Carmona, P. FT-IR study of multilamellar lipid dispersions containing cholesteryl linoleate and dipalmitoylphosphatidylcholine. *Chem. Phys. Lipids* **1999**, *97*, 119–128. [CrossRef]

46. Herrero, A.M.; Carmona, P.; Pintado, T.; Jiménez-Colmenero, F.; Ruíz-Capillas, C. Infrared spectroscopic analysis of structural features and interactions in olive oil-in-water emulsions stabilized with soy protein. *Food Res. Int.* **2011**, *44*, 360–366. [CrossRef]

47. Carmona, P.; Ruiz-Capillas, C.; Jiménez-Colmenero, F.; Pintado, T.; Herrero, A.M. Infrared study of structural characteristics of frankfurters formulated with olive oil-in-water emulsions stabilized with casein as pork backfat replacer. *J. Agric. Food Chem.* **2011**, *59*, 12998–13003. [CrossRef] [PubMed]

48. Herrero, A.M.; Carmona, P.; Cofrades, S.; Jiménez-Colmenero, F. Raman spectroscopic determination of structural changes in meat batters upon soy protein addition and heat treatment. *Food Res. Int.* **2008**, *41*, 765–772. [CrossRef]

49. Kurt, A.; Gençcelep, H. Enrichment of meat emulsion with mushroom (Agaricus bisporus) powder: Impact on rheological and structural characteristics. *J. Food Eng.* **2018**, *237*, 128–136. [CrossRef]

Article

Strategy towards Replacing Pork Backfat with a Linseed Oleogel in Frankfurter Sausages and Its Evaluation on Physicochemical, Nutritional, and Sensory Characteristics

Daniel Franco [1], Artur J. Martins [2], María López-Pedrouso [3], Laura Purriños [1], Miguel A. Cerqueira [4], António A. Vicente [2], Lorenzo M. Pastrana [4], Carlos Zapata [3] and José M. Lorenzo [1,*]

1   Centro Tecnológico de la Carne de Galicia, Rúa Galicia Nº 4, Parque Tecnológico de Galicia, San Cibrán das Viñas, 32900 Ourense, Spain
2   Centre of Biological Engineering, University of Minho, Campus de Gualtar, 4710-057 Braga, Portugal
3   Department of Zoology, Genetics and Physical Anthropology, University of Santiago de Compostela, 15872 Santiago de Compostela, Spain
4   International Iberian Nanotechnology Laboratory, Av. Mestre José Veiga s/n, 4715-330 Braga, Portugal
*   Correspondence: jmlorenzo@ceteca.net; Tel.: +34-988548277

Received: 3 July 2019; Accepted: 22 August 2019; Published: 26 August 2019

Abstract: Different health institutions from western countries ha–ve recommended a diet higher in polyunsaturated fats, especially of the *n*-3 family. However, this is not a trivial task, especially for meat-processing sectors. The objective of this work was to assess the influence of replacing pork backfat with linseed oleogel on the main quality parameters of frankfurters. The frankfurters were formulated by the pork backfat replacement of 0% (control), 25% (SF-25), and 50% (SF-50), using a linseed oleogel gelled with beeswax. The determination of quality parameters (pH, colour, chemical composition, and texture parameters), the fatty acid profile, and the sensory evaluation was carried out for each batch. The fatty acid profile was substantially improved, and the saturated fatty acid (SFA) content was reduced from 35.15g/100g in control sausages to 33.95 and 32.34g/100 g in SF-25 and SF-50, respectively, and more balanced ratios *n*-6/*n*-3 were achieved. In addition, the sausages with linseed oleogel also decreased the cholesterol content from 25.08 mg/100 g in control sausages to 20.12 and 17.23 mg/100 g in SF-25 and SF-50, respectively. It may therefore be concluded that these innovative meat products are a healthier alternative. However, sensory parameters should be improved in order to increase consumer acceptability, and further research is needed.

Keywords: low fat; beeswax; cholesterol; saturated fatty acid (SFA); *n*-6/*n*-3; preference test

1. Introduction

The consumer demand for healthy products is expected to grow in the coming years, frankfurter sausages being one of the most popular, with a large market worldwide for their convenience and price. Even though their intrinsic characteristics can vary significantly, pork frankfurters may contain up to 23% fat and 8.7% saturated fatty acids (SFA) [1], which converts them into less attractive meat products. It is a well-known fact that SFA and trans fatty acids (TFA) provide a suitable texture and juiciness in meat processed products, but they have detrimental effects on human health, such as cardiovascular disease or metabolic syndrome [2]. For this reason, the WHO recommends reducing the energy intake of total fats to less than 30% of the total daily diet and preventing SFA and TFA in the diet [3]. As animal fats are richer in SFA and TFA than those from a vegetal source, the partial replacement has become a strategic target for the meat industry in order to develop healthier meat products [4–8]. Indeed, vegetable oils in sausage formulations have been used for the production of low/improved fat frankfurters [9,10]. However, since vegetable oils are liquid at room temperature, this constitutes a major problem, resulting in substantial differences in texture, color, and flavor with regard to beef or pork backfat.

Trying to address these issues, food researchers have shown an increased interest in edible oleogels. The formation of oleogels requires an oleogelator that will form a network allowing the conversion of liquid oil to a solid substance, and it may also include bioactive compounds [11,12]. The properties of the oleogels are affected by the type, concentrations, and crystallization temperature of the oleogelator, by the oil medium and presence of other additives [2]. Oleogels based on canola oil and ethylcellulose as the organogelator have proven to be an alternative lipid phase in comminuted meat products [13,14], meat batters [15], and breakfast sausages [16]. However, temperatures required to dissolve ethylcellulose in oil are above 130 °C, which makes the whole process more complex. Alternatively, beeswax is certainly regarded as one of the most commercially valuable waxes, used to wrap cheese or as a food additive, which also can be used as an oleogelator. Beeswax was evaluated with different types of oils [11,17], but there is still insufficient data in meat processed products with few exceptions, such as beef burgers [18] or pâtés [19]. On the other hand, linseed oil is widely used in functional food because of its high level of α-linolenic (~55%), which is converted into long chain n-3 fatty acids (FA). Moreover, Fayaz et al. [20] have reported that textural properties of oleogels composed by beeswax and linseed oil were stable, showing greater firmness and stickiness than those from other oilseeds (canola, sesame, and sunflower). In addition, meat products enriched with linseed oil enhance the FA profile in health terms. Recently, Gómez–Estaca et al. [19] tested a mixture of olive, linseed, and fish oil together with beeswax to replace animal fat in pork liver pâtés, concluding that stability, texture, color, and sensory attributes were not significantly affected.

Therefore, the aim of this research was to study the effect of replacing pork backfat with different substitution levels with oleogels, elaborated using linseed oil and beeswax, on the main quality attributes of frankfurters (color, texture, and sensory properties).

2. Materials and Methods

2.1. Oleogel Elaboration

For the production of beeswax-based oleogel, a commercial linseed oil Vitaquell® with 72% polyunsaturated (approx. 55% of α-linolenic), 19% monounsaturated, and 9% saturated FAs was used as the oil phase. Oleogels with 8% (*w/w*) gelator were produced for all the fat replacement experiments. To ensure the solubilization of beeswax in linseed oil, the gelator was dispersed under stirring at 80 °C (above wax melting point) for at least 30 min. After that period of time, the oleogels were left cooling at ambient temperature until gel formation.

2.2. Frankfurter Production

Three different batches (1.5 kg per batch) were produced: A control batch (CO), elaborated only with pork backfat, two batches of sausage frankfurter (SF), SF-25, and SF-50 with 25% and 50% of pork backfat replaced by oleogel, respectively. The amount of pork backfat added was 172.5 g, 129.4 g, and 86.25 g for the CO, SF-25 and SF-50, respectively, and the amount of linseed oleogel added was 43.1 g and 86.25 g for SF-25 and SF-50, respectively. These formulations were selected according to previous works developed in our lab. The other components were added in the same proportion: Pork jowl (438.8 g), pork lean (198.8 g), pork heart (240 g), water (172.5 g), ice (418.1 g), sodium caseinate (18.8 g), salt (11.2 g), and commercial mix (189.4 g). The commercial mix (Ceylamix PT-F, Laboratorios Ceylamix, Valencia, Spain) consisted of potato starch, salt, milk, and soy protein, polyphosphates (E-450i, E-452i), dextrose glutamate monosodium (E-621), sodium ascorbate (E-301), sodium, nitrite (E-250), and paprika extract (E-160c).

The meat batter was elaborated as follows. The sodium caseinate and water at 60 °C were homogenized in a Ultraturrax T25 basic (IKA-Werke, Staufen, Germany) for 2 min at a ratio of 5:1. Pork backfat or oleogel were then added and emulsified for 3 min. Once the emulsifying process was finished, the mixture was refrigerated at ambient temperature. Meat batters were elaborated by chopping the pork lean, jowl, and heart in a cutter (Cutter K30, Talsa, Talsabell S.A., Valencia, Spain) at a low speed for 1 min. Afterwards, the pork backfat or oleogel mass was added and chopped for 1 additional min. Finally, the commercial mix was included in the meat batter and mixed for 5 min. During this process, the overall temperature did not exceed 8 °C. After emulsification, each batch of meat batter was stuffed into 25 mm collagen casings and divided into sausages of 8 cm of length.

The sausages were cooked in a water bath (Marmite Mera 120 × 70, Talsa, Talsabell S.A., Valencia, Spain) at 90 °C for 20 min. The sausages were then immersed in an ice water-bath for 5 min, placed in polyethylene bags, vacuum packaged, and pasteurized at 90 °C for 30 min. Finally, the sausages were transferred to a cooler at 2 °C and stored for 24 hours until analysis. A total of 30 sausages of 100 grams (five samples per each batch × three batches × two replicates) were analyzed for different quality traits. Before analysis, sausages were minced in a blender (Moulinex, Barcelona, Spain).

2.3. Physicochemical Analysis

2.3.1. Determination of Quality Parameters: pH, Color, and Chemical Composition

The pH value was measured using a portable pH-meter (Hanna Instruments, Eibar, Spain), equipped with a penetration probe. The color determination of sausages was assessed from fresh cut cross-sections in three different points with a portable colorimeter (Konica Minolta CM-600d, Osaka, Japan), according to CIELAB space [21]: Lightness, (L*); redness, (a*); yellowness (b*), with the next settings machine (pulsed xenon arc lamp, angle of 0° viewing angle geometry, and aperture size of 8 mm). Hue (H_{ab}) and chroma (C*) were calculated from the a* and b* values according to expressions:

$$C^* = \sqrt{(a^*)^2 + (b^*)^2} \ and \ h_{ab} = acr\tan\frac{b^*}{a^*} \tag{1}$$

Moisture, protein, and ash were assessed following the International Organization for Standardization (ISO) recommended standards [22–24]. Total fat was extracted according to the American Oil Chemists' Society (AOCS) official procedure [25], while carbohydrate contents were estimated by the difference. The quantification of total cholesterol, saponification, extraction, and identification was carried out with high performance liquid chromatography (HPLC) in the normal phase, according to Dominguez et al. [26].

2.3.2. Fatty Acid Profile

The total fat was extracted from 12.5 grams of the sample and 50 milligrams were utilized to determine the FA profile. Total FAs were transesterified, following Dominguez et al. [26]. The separation and quantification of the fatty acid methyl esters (FAMEs) was performed using a gas chromatograph (GC-Agilent 6890 N; Agilent Technologies Spain, S.L., Madrid, Spain) equipped with a flame ionization detector following the chromatographic conditions described by Dominguez et al. [26]. Data of FAME profiles were expressed in grams per 100 grams of fat.

To assess the nutritional properties of sausages, the ratios polyunsaturated fatty acid (PUFA)/SFA, *n*-6/*n*-3, hypocholesterolemic/Hypercholesterolemic (h/H) ratios, the nutritional value (NV), the indexes of atherogenicity (AI), and thrombogenicity (TI) were determined. The hypocholesterolemic/Hypercholesterolemic (h/H) ratio was calculated according to Santos–Silva et al. [27]: h/H = ((sum of C18:1*n*9c, C18:2*n*6, C18:3*n*6, C18:3*n*3, C20:3*n*6, C20:4*n*6, C20:5*n*3, and

C22:5n3)/(sum of C14:0 and C16:0)). The atherogenic index (AI) and thrombogenic index (TI) were calculated according to Ulbricht and Sauthgate [28]:

$$AI = [(4 \times C14{:}0) + C16{:}0]/[(\Sigma PUFA) + (\Sigma MUFA)]; \qquad (2)$$

$$TI = [C14{:}0 + C16{:}0 + C18{:}0]/[(0.5 \times \Sigma MUFA) + (0.5 \times n6) + (3 \times n3) + (n3/n6)].$$

2.3.3. Texture Profile Analysis

Sausage pieces of 1 × 1 × 2.5 cm (height × width × length) were compressed at a crosshead speed of 3.33 mm/s in a texture analyzer (TA.XTplus, Stable Micro Systems, Vienna Court, UK). The following parameters were recorded: Hardness, cohesiveness, springiness, gumminess, chewiness, and adhesiveness by compressing to 80%, using a compression probe with 19.85 cm^2 of surface contact.

2.3.4. Sensory Evaluation

A pilot consumer test was performed by 32 panelists, recruited among personnel of the Meat Technology Center. The testing was carried out in a room equipped with individual tasting booths under white light [29]. Water and bread without salt were available for rinsing the mouth between samples. Samples were cooked in a hot steam oven (CombiMaster®Plus, Rational, Landsberg am Lech, Germany) until a core temperature of 70 ± 2 °C was reached. The temperature was controlled using an oven thermometer inserted lengthwise into one of the sausages. After cooking, the sausages were sliced into 1.5 cm thick pieces and served hot. The serving temperature of the samples was 50 ± 2 °C. The three samples were presented to the panelists following a balanced order for each of them [30]. Panelists were asked to rank the samples according to their own preference related to appearance, odor, hardness, juiciness, taste, and global perception. They also evaluated the overall acceptability of each sample using a 7-point hedonic scale, varying from (1) "liked very much" to (7) "disliked very much", according to Meilgaard et al. [31].

2.4. Statistical Analysis

All statistical analysis was performed using SPSS v. 23 package (IBM SPSS, Chicago, Illinois, USA). The effect of inclusion of oleogel at two levels (25 and 50%) was evaluated employing a mixed-model analysis of variance (ANOVA), where these traits were set as dependent variables, oleogel concentration (0, 25, and 50%) as a fixed effect, and a replicate as a random effect. The pairwise differences between least-square means were evaluated by Duncan's test. Correlations among traits were determined by Pearson's linear correlation coefficient.

The XLSTAT-Sensory version 2018 software was used to analyze all the sensory data. To analyze differences in the sensorial parameters evaluated by the panelist preference ranking test, the Friedman rank sum test was performed [32], using a significance level of 95% to determine whether the panelists were able to discriminate among samples. The least significant difference was used to determine whether significant differences ($p < 0.05$) existed among sausages.

Overall acceptability data from the evaluation was analyzed by means of a linear mixed model. A sample was specified as a fixed effect while a panelist was specified as a random effect. A significance level of 95% was considered in the analysis and Tukey's test was used to separate means.

3. Results

3.1. Chemical Composition of Frankfurters

The statistical analysis showed significant differences for all evaluated parameters; however, the numerical values among batches were similar, with the exception of cholesterol content (Table 1). These outcomes can be explained by the fact that the amount of ingredients in all formulations was in fact the same, except the subcutaneous fat from pork which was replaced by the linseed oleogel. These slight observed differences could be attributed to the lack of homogeneity (backfat, jowl, lean, and heart) in the frankfurter elaboration procedure.

Table 1. Chemical composition of a frankfurter, expressed in percentage (mean values ± standard deviation of 10 samples).

Parameter	Control	SF-25	SF-50	SEM	*p*-Value
Moisture (%)	57.55 ± 1.02 [a]	55.98 ± 0.49 [c]	56.74 ± 0.67[b]	0.17	$p < 0.001$
Protein (%)	11.19 ± 0.33 [a]	10.63 ± 0.39 [b]	10.62 ± 0.22[b]	0.07	$p = 0.001$
Ash (%)	4.24 ± 0.17 [a]	4.05 ± 0.09 [b]	4.04 ± 0.10[b]	0.03	$p = 0.002$
Fat (%)	18.35 ± 0.60 [b]	20.03 ± 1.04 [a]	18.95 ± 0.72[b]	0.19	$p < 0.001$
Carbohydrate (%)	8.65 ± 0.57 [b]	9.30 ± 0.66 [a]	9.63 ± 0.60[a]	0.13	$p = 0.005$
Cholesterol (mg/100 g)	25.08 ± 0.88 [a]	20.12 ± 0.77 [b]	17.23 ± 0.38[c]	0.87	$p < 0.001$

Saturated fatty (SF), Standard error of mean (SEM), [a–c] Means in the same row with different letters differing significantly ($p < 0.05$).

Differences in moisture content ranged between 55.55–56.74% for all the batches, while the protein percentage varied from 10.62 to 11.19%, resulting in slightly higher values for the control samples. In addition, the replacement with linseed oleogel had a significant ($p < 0.05$) effect on protein content. Ash content was significantly affected by replacement of pork backfat, ranging between 4.04 and 4.24%. Regarding the fat content, the differences among batches were small (18.35–20.03%), although they reached statistical ($p < 0.05$) relevance and the cholesterol amount varied significantly ($p < 0.001$) among batches, with the highest content (25.08 mg/100g) in the control sausages.

3.2. Fatty Acid Profile of Frankfurters

The influence of the partial replacement of pork backfat at two levels by linseed oleogel on the FA profile is shown in Table 2. As expected, this partial substitution caused a significant effect in all FAs with the exception of α-linoleic acid (C18:2n6c). This is due to the healthy FA profile of the linseed oleogel. However, the most predominant FAs in all batches were monosaturated fatty acid (MUFA), followed by SFA and polyunsaturated fatty acid (PUFA).

Regarding individual FA, oleic acid (C18:1n9c) was the most considerable FA followed by palmitic acid (C16:0), α-linoleic acid, and stearic acid (C18:0). The replacement of pork backfat by linseed oleogel reduced the SFA content from 35.15 g/100 g obtained in control sausages to 33.95 and 32.34 g/100 g in SF-25 and SF-50, respectively. The most predominant SFA was palmitic acid (C16:0), followed by stearic acid (C18:0). The palmitic content decreased significantly ($p < 0.05$) from 22.23 to 20.07 g/100 g for the control and SF-50 treatment, respectively. Moreover, stearic content showed a significant reduction from 10.92 to 10.37 g/100 g. The MUFA content was also affected by linseed replacement, presenting the highest values in the control sausages (48.06 mg/100 g). Differences are mainly due to oleic content variation along the three batches. In this sense, linseed seed is poor in this FA, causing a significant decrease in the final formulation.

Concerning PUFA, their content was significantly ($p < 0.05$) affected by linseed inclusion. Indeed, the PUFA content increased from 16.77 mg/100g for the control to 20.10 and 25.46 mg/100 g for SF-25 and SF-50, respectively. The linoleic acid did not show significant ($p > 0.05$) differences among batches; however, variations in PUFA content are directly associated to the linolenic content. This was a desired consequence, which had immediate consequences on n-3 PUFA content as well as in PUFA/SFA and n-6/n-3 ratios. Indeed, control sausages had the lowest PUFA/SFA ratio (0.47) and the higher n-6/n-3 (14.92). Thus, these ratios could be significantly ($p < 0.05$) increased to 0.78 and decreased to 1.61, respectively, with replacement of 50% of pork backfat by linseed oleogel.

Table 2. Fatty acid profile (g/100 g of fat) of frankfurter (mean values ± standard deviation of 10 samples).

Fatty Acid	Control	SF-25	SF-50	SEM	*p*-Value
C14:0	1.26 ± 0.01 [a]	1.27 ± 0.02 [a]	1.16 ± 0.03 [b]	0.01	$p < 0.001$
C16:0	22.23 ± 0.02 [a]	21.49 ± 0.25 [b]	20.07 ± 0.40 [c]	0.17	$p = 0.002$
C16:1*n*7	2.27 ± 0.07 [a]	2.28 ± 0.01 [a]	1.95 ± 0.09 [b]	0.03	$p < 0.001$
C17:0	0.38 ± 0.01 [a]	0.35 ± 0.01 [b]	0.31 ± 0.01 [c]	0.005	$p < 0.001$
C18:0	10.92 ± 0.04 [a]	10.47 ± 0.16 [b]	10.37 ± 0.03 [c]	0.04	$p < 0.001$
C18:1*n*7t	0.30 ± 0.01 [a]	0.28 ± 0.01 [b]	0.24 ± 0.01 [c]	0.04	$p < 0.001$
C18:1*n*9c	40.93 ± 0.04 [a]	39.00 ± 0.59 [b]	36.13 ± 0.78 [c]	0.37	$p < 0.001$
C18:1*n*7c	3.20 ± 0.01 [a]	3.10 ± 0.03 [b]	2.75 ± 0.10 [c]	0.036	$p < 0.001$
C18:2*n*6c	14.31 ± 0.04	14.30 ± 0.06	14.34 ± 0.02	0.008	$p = 0.965$
C:20:0	0.15 ± 0.01 [c]	0.16 ± 0.01 [b]	0.17 ± 0.0 [a]	0.002	$p < 0.001$
C18:3*n*3	0.80 ± 0.03 [c]	4.27 ± 1.09 [b]	9.68 ± 1.49 [a]	0.70	$p < 0.001$
C20:2*n*6	0.68 ± 0.01 [a]	0.62 ± 0.01 [b]	0.54 ± 0.02 [c]	0.01	$p < 0.001$
C20:4*n*6	0.46 ± 0.01 [a]	0.41 ± 0.01 [b]	0.40 ± 0.01 [c]	0.005	$p < 0.001$
C22:5*n*3	0.09 ± 0.01 [a]	0.08 ± 0.01 [b]	0.07 ± 0.01 [c]	0.001	$p < 0.001$
SFA	35.15 ± 0.07 [a]	33.95 ± 0.41 [b]	32.34 ± 0.45 [c]	0.22	$p < 0.001$
MUFA	48.06 ± 0.05 [a]	45.93 ± 0.65 [b]	42.18 ± 1.03 [c]	0.46	$p < 0.001$
PUFA	16.77 ± 0.07 [c]	20.10 ± 1.06 [b]	25.46 ± 1.48 [a]	0.69	$p < 0.001$
*n*6/*n*3	14.92 ± 0.44 [a]	3.87 ± 2.01 [b]	1.61 ± 0.39 [c]	1.10	$p < 0.001$
PUFA/SFA	0.47 ± 0.01 [c]	0.59 ± 0.04 [b]	0.78 ± 0.05 [a]	0.02	$p < 0.001$
IA	0.42 ± 0.01 [a]	0.40 ± 0.01 [b]	0.36 ± 0.01 [c]	0.004	$p < 0.001$
IT	0.98 ± 0.01 [a]	0.75 ± 0.07 [b]	0.53 ± 0.05 [c]	0.03	$p < 0.001$
h/H	2.53 ± 0.01 [c]	2.67 ± 0.04 [b]	2.97 ± 0.08 [a]	0.03	$p < 0.001$

Standard error of mean (SEM), the saturated fatty acid (SFA), monounsaturated fatty acid (MUFA), polyunsaturated fatty acid (PUFA), atherogenicity (IA), thrombogenicity (IT), [a–c] Means in the same row with different letters differing significantly ($p < 0.05$).

Finally, replacing pork backfat with a linseed oleogel also had a significant effect ($p < 0.001$) on the IA and IT and on h/H. Sausages with a replacement of 25% and 50% obtained the lowest values for IA and IT (0.42 vs. 0.40 vs. 0.36; $p < 0.001$ for control, SF-25 and SF-50, respectively, in the case of IA and 0.98 vs. 0.75 vs. 0.53; $p < 0.001$ for control, SF-25 and SF-50, respectively), showing the better nutritional fatty acid profile. Regarding the h/H ratio, sausages replaced with 50% of lineseed oleogel obtained the highest values, with the highest percentage of hypocholesterolemic FA, α-linolenic, and the lowest amounts of hypercholesterolemic FA (C14:0 and C16:0). On the contrary, control sausages displayed an opposite trend in respect to the amounts of hypocholesterolemic and hypercholesterolemic FA, resulting in the lowest h/H values. However, for all sausage batches studied this ratio was higher than 2.5, which is considered as favorable.

3.3. Quality Parameters: pH and Color Assessment of Frankfurters

The pH values were significantly ($p < 0.001$) influenced by replacement with linseed oleogel, although numerical values were very similar. Regarding color parameters (L*, a*, and b*), they were significantly ($p < 0.001$) influenced by substitution of pork backfat with linseed oleogel. Specifically, luminosity and yellowness increase with the amount of linseed oleogel (L* from 61.47 to 69.37 and b* from 16.61 to 18.85 for the control and SF-50, respectively). On the contrary, redness value decreased from 12.50 to 9.06 for control and SF-50, respectively (Figure 1).

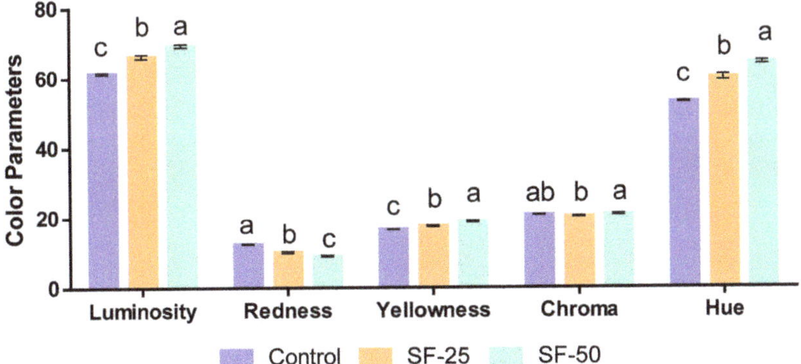

Figure 1. Color parameters of frankfurters, [a–c] Different letters indicate statistically significant differences ($p < 0.05$), (mean values ± standard deviation of 10 samples).

3.4. Texture Profile Analysis of Frankfurters

The textural profile analysis shows that the replacement of pork backfat with linseed oleogel led to significant variations in the following textural parameters: Hardness, cohesiveness, gumminess, and chewiness (Figure 2). Cohesiveness, gumminess, and chewiness significantly increased in SF-50, with respect to control sausages, but the differences were relatively slight.

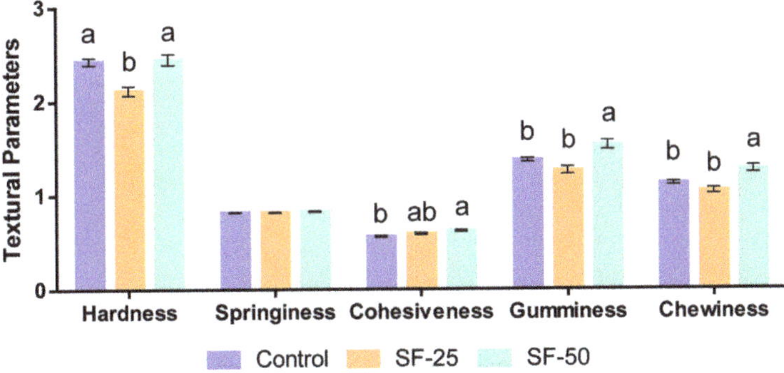

Figure 2. Textural parameters of frankfurters. Hardness (Kg), Springiness (mm), Gumminess (Kg), and Chewiness (kg mm), [a–c] Different letters indicate statistically significant differences ($p < 0.05$), (mean values ± standard deviation of 10 samples).

3.5. Sensory Attributes of Frankfurters

Sensory analysis indicated that there were significant differences in the preference ($p < 0.05$), influenced by the replacement of pork backfat with linseed oleogel (Table 3). The appearance is significant to consumer preference and acceptability of products. Panelists showed a clear preference ($p < 0.05$) for control samples over SF-25 and SF-50 samples. Regarding color parameters, the control sample obtained the lowest yellowness value, suggesting a higher acceptability. In meat products, the yellow color is associated with rancid foods caused by lipid oxidation. A significant difference ($p < 0.05$) was detected between the control and the remaining samples for the odor and taste. The control sample was the most preferred in both cases. Hardness only showed a slightly higher value in the 25% substitution sausage without significant influence in the evaluation for ranking preference. In the case of juiciness, a significantly lower score ($p < 0.05$) showed for the SF-50, whereas no differences were detected between the control and SF-25. Finally, control samples obtained higher scores for global perception compared to other samples (86 vs. 58 and 48 for control, SF-25, and SF-50 batches, respectively).

Table 3. Preference data: Rank sums. Ranking preference: 1 the lowest, 3 the highest preferred.

Parameter	Control	SF-25	SF-50
Appearance	84 [a]	44 [b]	64 [b]
Odor	86 [a]	56 [b]	50 [b]
Hardness	60	74	58
Juiciness	72 [a]	72 [a]	48 [b]
Taste	82 [a]	60 [b]	50 [b]
Global Perception	86 [a]	58 [b]	48 [b]

[a–b] Different letters indicate statistically significant differences ($p < 0.05$).

4. Discussion

Among batches, variations in moisture content were slightly lower than some described by other authors on frankfurters [10,33,34]. Differences in protein content between control sausages and the other batches (SF-25 and SF-50) could be explained by the presence of small lean portions in pork backfat. Other authors have reported frankfurter sausages with protein content between 12.5–15% [10,33,34]. The range of ashes content was higher than those shown by other authors [10,33,34], who indicated values in the range 2.58 to 3.39%. These differences could be explained by variations in the amount of sodium caseinate, salt, and commercial mix among recipes of the studies.

Previous studies have noted the importance of animal fat in this type of meat products, because it plays a significant role in flavor intensity, juiciness, and tenderness. In comminuted meat products and frankfurters, values of around 26% have been found [13,35], but other lower values have been reported as 20.8% in breakfast sausage [16], 14.41% in sausages elaborated with microencapsulated fish oil [34], or even 10.3% for low-fat sausages elaborated with Konjac gel [10]. Regarding cholesterol content, it is remarkable that frankfurters formulated with linseed had the lowest cholesterol content with respect to the control sausages (25.08 vs. 17.23; $p < 0.05$), despite the fact that this result was expectable, because cholesterol is inherent to animal tissues, hence it should not be found in linseed. This is an important improvement from a nutritional point of view.

Overall, the individual FA in the following order (oleic > palmitic > linoleic > stearic) match those reported in earlier studies of frankfurters [33,34,36]. It should be noted that a noticeable SFA reduction (in the range 3.4–7.99%) was achieved as a result of the replacement of pork backfat by linseed oleogel. A similar reduction range (3–8%) was obtained with sausages containing fish oil [37]. Other authors have reported higher reductions in SFA content when replacing pork backfat with fish and/or vegetable oils, e.g., 4.5–11.8% [34] and 3–25% [38] in frankfurter sausages.

Reductions in the main SFA (palmitic and stearic acids) have important and beneficial effects on human health in accordance with the recommendations of WHO [3]. Recommendations include reducing SFA intake, due to the raise of low-density lipoprotein (LDL)-cholesterol producing atherogenic and hypercholesterolemic effects [39]. This result may be explained by the fact that linseed has a lower SFA percentage than pork backfat. These findings match those mentioned in previous reports, in regard to the fat substitution in frankfurters with vegetable oils [10,33,34].

On the other hand, lipid oxidation affects color, texture, nutritional value, taste, and aroma leading to rancidity, which is responsible for off-flavors and unacceptable taste, which are important reasons for consumer rejection [40,41]. Lipid oxidation is a rather complex process, in which unsaturated fatty acids react with molecular oxygen via free radical chain-forming peroxides [42]. The first auto-oxidation is followed by a series of secondary reactions, which lead to lipid degradation and the development of oxidative rancidity products. However, lipid oxidation (Thiobarbituric acid reactive substances values) were not determined in our sausage samples.

Variations in PUFA content, as well as PUFA/SFA and *n*-6/*n*-3 ratios, have important implications for the development of healthy meat products, as it is well known that balanced ratios of *n*-6/*n*-3 and PUFA/SFA have positive effects on human health. Effectively, excessive *n*-6 PUFA content and therefore greater *n*-6/*n*-3 ratios can result in diseases, such as cardiovascular pathology and prostate cancer, whereas an excessive increase of *n*-3 PUFA exerts suppressive effects [43]. However, western diets are increasing the amount of *n*-6 PUFA. Higher proportions of *n*-3 PUFA have been recommended by European Food Safety Authority (EFSA) [39] and Food and Agriculture Organization (FAO) [44] in order to decrease the *n*-6/*n*-3 ratio for the prevention of above-mentioned diseases. The present results agree with the findings of other studies in which the inclusion of olive and fish oils, mixture or alone, diminished the *n*-6/*n*-3 ratio in frankfurters [34] and beef burgers [45].

The color parameters of batches formulated with linseed oleogel are strongly related to color characteristics of raw ingredients used in the formulation. The increase in luminosity and yellowness may be explained by the yellow color of linseed oleogel and is associated with the amount used in the elaboration. Other authors have reported similar results in different meat products, replacing pork backfat with vegetable oils (olive, canola), such as frankfurters [34], breakfast sausages [16], pate [46], or pork patties [47]. On the contrary, Barbut et al. [16] indicated a significant decrease in the luminosity of breakfast sausages elaborated with canola oleogel. The significant ($p < 0.001$) reduction of redness values in SF-50 (9.06) and SF-25 (10.15), with respect to the control samples (12.50), seems to be consistent with the study of Lopez-Lopez et al. [9], which found that the substitution of pork backfat by olive oil decreased the redness value in frankfurters.

One of the main issues in the reformulation of meat products, such as frankfurters or sausages, is regarding textural properties, because some of these properties were quite different in new products than original products. For instance, Barbut et al. [16] indicated that the replacement of beef fat by canola oil in comminuted products led to a firmer and higher rubbery product, which is unsatisfactory. However, in our study, changes in the sausage textural profile were not remarkable, in agreement with [33], who observed no influence on the hardness or chewiness of frankfurters with olive oil replacing pork backfat. However, this issue is controversial, as the increase of hardness and firmness in frankfurters with olive oil instead of pork backfat [48,49], as well as no influence [50], has also been reported. In addition, in the present study, a negative correlation between hardness and fat content was found ($r = -0.560$; $p < 0.01$). A similar behavior was noticed for gumminess and chewiness ($r = -0.444$ and -0.428, $p < 0.05$, respectively) while adhesiveness was positive, correlated with total fat content ($r = 0.414$, $p < 0.05$).

Modifications in the formulation resulted in global perception being significantly different among samples. Thus, samples with linseed oleogel were assessed with a lower preference. The overall acceptability for sausages depends on many attributes and their interactions. Panelists showed significant differences regarding the sausages with substitution of pork backfat, with overall acceptability scores of 2.06 ± 0.17, 2.69 ± 0.19 and 3.19 ± 0.17 for CO, SF-25, and SF-50. Although the sausage control showed more acceptability than the modified sausages, the three formulations scored in the positive part of the hedonic scale used in this study, obtaining a positive acceptance. Barbut et al. [13] noticed that there may be a hardness and juiciness tipping-point based on the ethylcellulose (EC) concentration in the organogel, where frankfurters containing organogels of lower EC content resembled those prepared with beef fat (control group). In addition, they reported that organogels with higher EC concentration-yielded frankfurters that resembled the control group, potentially allowing for the custom formulation of hardness and other sensory characteristics. Organogel samples formulated with sorbitan monostearate can further help tailor textural and sensory characteristics of emulsion type meat products by adding plasticity to the organogel structure [13].

5. Conclusions

Linseed oleogel was successfully inserted into frankfurters in order to replace the pork backfat. The FA profile improved with the use of oleogels, the values of SFA content and the *n*-6/*n*-3 ratio reduced significantly, and the cholesterol amount also decreased significantly in the sausages with linseed oleogel. These healthier newly emerging meat products are being screened in the meat industry. However, other sensory parameters could not be substantially improved, such as color, which increased the yellowness with linseed oleogel. Moreover, cohesiveness, gumminess, and chewiness presented higher values in this new product. More research is needed to enhance the sensory properties and lipid oxidation in precooked products with linseed oleogels.

Author Contributions: D.F., M.A.C., L.M.P., A.A.V., C.Z., and J.M.L. conceived the idea and participated with all authors in drafting the manuscript. A.J.M, M.L.P., and L.P. elaborated the samples and analyzed it. They also participated in the data interpretation, review, and editing of the manuscript.

Funding: Artur Martins is the recipient of a fellowship supported by a doctoral advanced training (call NORTE-69-2015-15) funded by the European Social Fund under the scope of Norte2020—Programa Operacional Regional do Norte. Jose M. Lorenzo is a member of the HealthyMeat network, funded by CYTED (ref. 119RT0568). Thanks to GAIN (Axencia Galega de Innovación) for supporting this research (grant number IN607A2019/01).

Conflicts of Interest: The authors declare no conflict of interest.

References

1. USDA. *National Nutrient Database for Standard Reference*; Basic Report: 07939, Frankfurter, Meat; U.S. Department of Agriculture: Washington, DC, USA, 2019.

2. Jiménez-Colmenero, F.; Salcedo-Sandoval, L.; Bou, R.; Cofrades, S.; Herrero, A.M.; Ruiz-Capillas, C. Novel applications of oil-structuring methods as a strategy to improve the fat content of meat products. *Trends Food Sci. Technol.* **2015**, *44*, 177–188. [CrossRef]

3. WHO. 2018. Available online: https://www.who.int/news-room/fact-sheets/detail/healthy-diet (accessed on 15 June 2019).

4. De Carvalho, F.A.L.; Lorenzo, J.M.; Pateiro, M.; Bermúdez, R.; Purriños, L.; Trindade, M.A. Effect of guarana (Paullinia cupana) seed and pitanga (Eugenia uniflora L.) leaf extracts on lamb burgers with fat replacement by chia oil emulsion during shelf life storage at 2 °C. *Food Res. Int.* **2019**, *125*, 108554. [CrossRef]

5. Heck, R.T.; Saldaña, E.; Lorenzo, J.M.; Correa, L.P.; Fagundes, M.B.; Cichoski, A.J.; de Menezes, C.R.; Wagner, R.; Campagnol, P.C.B. Hydrogelled emulsion from chia and linseed oils: A promising strategy to produce low-fat burgers with a healthier lipid profile. *Meat Sci.* **2019**, *156*, 174–182. [CrossRef] [PubMed]

6. Da Silva, S.L.; Amaral, J.; Ribeiro, M.; Sebastiao, E.; Vargas, C.; Frazen, F.; Schneider, G.; Lorenzo, J.M.; Fries, L.L.M.; Cichoski, A.J.; et al. Fat replacement by oleogel rich in oleic acid and its impact on the technological, nutritional, oxidative, and sensory properties of Bologna-type sausages. *Meat Sci.* **2019**, *149*, 141–148. [CrossRef] [PubMed]

7. Munekata, P.E.S.; Dominguez, R.; Campagnol, P.C.B.; Franco, D.; Trindade, M.A.; Lorenzo, J.M. Effect of natural antioxidants on physicochemical properties and lipid stability of pork liver pâté manufactured with healthy oils during refrigerated storage. *J. Food Sci. Technol.* **2017**, *54*, 4324–4334. [CrossRef] [PubMed]

8. Heck, R.T.; Guidetti, R.; Etchepare, M.A.; dos Santos, L.A.A.; Cichoski, A.J.; Ragagnin, C.; Smanioto, J.; Lorenzo, J.M.; Wagner, R.; Campagnol, P.C.B. Is it possible to produce a low-fat burger with a healthy n-6/n-3 PUFA ratio without affecting the technological and sensory properties? *Meat Sci.* **2017**, *130*, 16–25. [CrossRef] [PubMed]

9. López-López, I.; Cofrades, S.; Jiménez-Colmenero, F. Low-fat frankfurters enriched with n-3 PUFA and edible seaweed: Effects of olive oil and chilled storage on physicochemical, sensory and microbial characteristics. *Meat Sci.* **2009**, *83*, 148–154. [CrossRef] [PubMed]

10. Salcedo-Sandoval, L.; Cofrades, S.; Pérez, C.R.C.; Solas, M.T.; Jiménez-Colmenero, F. Healthier oils stabilized in konjac matrix as fat replacers in n-3 PUFA enriched frankfurters. *Meat Sci.* **2013**, *93*, 757–766. [CrossRef]

11. Patel, A.R.; Dewettinck, K. Edible oil structuring: An overview and recent updates. *Food Funct.* **2016**, *7*, 20–29. [CrossRef]

12. Martins, A.J.; Vicente, A.A.; Cunha, R.L.; Cerqueira, M.A. Edible oleogels: An opportunity for fat replacement in foods. *Food Funct.* **2018**, *9*, 758–773. [CrossRef]

13. Barbut, S.; Wood, J.; Marangoni, A. Potential use of organogels to replace animal fat in comminuted meat products. *Meat Sci.* **2016**, *122*, 155–162. [CrossRef]

14. Agregán, R.; Barba, F.J.; Gavahian, M.; Franco, D.; Khaneghah, A.M.; Carballo, J.; Ferreira, I.C.F.R.; Barreto, A.C.S.; Lorenzo, J.M. Fucus vesiculosus extracts as natural antioxidants for improving the physicochemical properties and shelf life of pork patties formulated with oleogels. *J. Sci. Food Agric.* **2019**, *99*, 4561–4570.

15. Zetzl, A.K.; Marangoni, A.G.; Barbut, S. Mechanical properties of ethylcellulose oleogels and their potential for saturated fat reduction in frankfurters. *Food Funct.* **2012**, *3*, 327–337. [CrossRef]

16. Barbut, S.; Wood, J.; Marangoni, A. Quality effects of using organogels in breakfast sausage. *Meat Sci.* **2016**, *122*, 84–89. [CrossRef]

17. Martins, A.J.; Cerqueira, M.A.; Fasolin, L.H.; Cunha, R.L.; Vicente, A.A. Beeswax organogels: Influence of gelator concentration and oil type in the gelation process. *Food Res. Int.* **2016**, *84*, 170–179. [CrossRef]

18. Moghtadaei, M.; Soltanizadeh, N.; Goli, S.A.H. Production of sesame oil oleogels based on beeswax and application as partial substitutes of animal fat in beef burger. *Food Res. Int.* **2018**, *108*, 368–377. [CrossRef]

19. Gómez-Estaca, J.; Herrero, A.M.; Herranz, B.; Álvarez, M.D.; Jiménez-Colmenero, F.; Cofrades, S. Characterization of ethyl cellulose and beeswax oleogels and their suitability as fat replacers in healthier lipid pâtés development. *Food Hydrocoll.* **2019**, *87*, 960–969. [CrossRef]

20. Fayaz, G.; Goli, S.A.H.; Kadivar, M. A Novel Propolis Wax Based Organogel: Effect of Oil Type on Its Formation, Crystal Structure and Thermal Properties. *J. Am. Oil Chem. Soc.* **2017**, *94*, 47–55. [CrossRef]

21. CIE. *International Commission on Illumination, Recommendations on Uniform Color Spaces, Color Difference Equations, Psychometric Color Terms*; Supplement Nr.15 to CIE publication Nr.15 (E-1.3.1) 1971/ (TO-1.3); Bureau Central de la CIE: Paris, France, 1978.

22. ISO 1442. *Determination of Moisture Content*; International Organization for Standardization: Geneva, Switzerland, 1997.

23. ISO 937. *Determination of Nitrogen Content*; International Organization for Standardization: Geneva, Switzerland, 1978.

24. ISO 936. *Determination of Ash Content*; International Organization for Standardization: Geneva, Switzerland, 1998.

25. AOCS. *AOCS Official Procedure Am 5-04. Rapid Determination of Oil/Fat Utilizing High Temperature Solvent Extraction*; American Oil Chemists Society: Urbana, IL, USA, 2005.

26. Dominguez, R.; Crecente, S.; Borrajo, P.; Agregán, R.; Lorenzo, J.M. Effect of slaughter age on foal carcass traits and meat quality. *Animal* **2015**, *9*, 1713–1720. [CrossRef]

27. Santos-Silva, J.; Bessa, R.J.B.; Santos-Silva, F. Effect of genotype, feeding system and slaughter weight on the quality of light lambs. II. Fatty acid composition of meat. *Livest. Prod. Sci.* **2002**, *77*, 187–194. [CrossRef]

28. Ulbricht, T.L.V.; Southgate, D.A.T. Coronary heart disease: Seven dietary factors. *Lancet* **1991**, *338*, 985–992. [CrossRef]

29. ISO 8589. *Sensory Analysis-General Guidance for the Design of Test Rooms*; International Organization for Standardization: Geneva, Switzerland, 2007.

30. Macfie, H.J.; Bratchell, N.; Greenhoff, K.; Vallis, L. Designs to balance the effect of order presentation and first-order carry-over effects in hall tests. *J. Sens. Stud.* **1989**, *4*, 129–148. [CrossRef]

31. Meilgaard, M.; Civille, G.V.; Carr, B.T. *Sensory Evaluation Techniques*; CRC Press: Boca Raton, FL, USA, 2006.

32. ISO 8587. *Sensory Analysis-Methodology-Ranking*; International Organization for Standardization: Geneva, Switzerland, 2006.

33. Delgado-Pando, G.; Cofrades, S.; Ruiz-Capillas, C.; Jiménez-Colmenero, F. Healthier lipid combination as functional ingredient influencing sensory and technological properties of low-fat frankfurters. *Eur. J. Lipid Sci. Technol.* **2010**, *112*, 859–870. [CrossRef]

34. Dominguez, R.; Pateiro, M.; Agregán, R.; Lorenzo, J.M. Effect of the partial replacement of pork backfat by microencapsulated fish oil or mixed fish and olive oil on the quality of frankfurter type sausage. *J. Food Sci. Technol.* **2017**, *54*, 26–37. [CrossRef]

35. Álvarez, D.; Delles, R.M.; Xiong, Y.L.; Castillo, M.; Payne, F.A.; Laencina, J. Influence of canola-olive oils, rice bran and walnut on functionality and emulsion stability of frankfurters. *LWT-Food Sci. Technol.* **2011**, *44*, 1435–1442. [CrossRef]

36. Ayo, J.; Carballo, J.; Serrano, J.; Olmedilla-Alonso, B.; Ruiz-Capillas, C.; Jiménez-Colmenero, F. Effect of total replacement of pork backfat with walnut on the nutritional profile of frankfurters. *Meat Sci.* **2007**, *77*, 173–181. [CrossRef]

37. Josquin, N.M.; Linssen, J.P.; Houben, J.H. Quality characteristics of Dutch-style fermented sausages manufactured with partial replacement of pork back-fat with pure, pre-emulsified or encapsulated fish oil. *Meat Sci.* **2012**, *90*, 81–86. [CrossRef]

38. Choi, Y.S.; Choi, J.H.; Han, D.J.; Kim, H.Y.; Lee, M.A.; Jeong, J.Y.; Kim, C.J. Effects of replacing pork back fat with vegetable oils and rice bran fiber on the quality of reduced-fat frankfurters. *Meat Sci.* **2010**, *84*, 557–563. [CrossRef]

39. EFSA. Scientific opinion on dietary reference values for fats, including saturated fatty acids, polyunsaturated fatty acids, monounsaturated fatty acids, trans fatty acids, and cholesterol. *EFSA J.* **2010**, *8*, 1461.

40. Domínguez, R.; Gómez, M.; Fonseca, S.; Lorenzo, J.M. Influence of thermal treatment on formation of volatile compounds, cooking loss and lipid oxidation in foal meat. *LWT-Food Sci. Technol.* **2014**, *58*, 439–445. [CrossRef]

41. Cossignani, L.; Giua, L.; Simonetti, M.S.; Blasi, F. Volatile compounds as indicators of conjugated and unconjugated linoleic acid thermal oxidation. *Eur. J. Lipid Sci. Technol.* **2014**, *116*, 407–412. [CrossRef]

42. Giua, L.; Blasi, F.; Simonetti, M.S.; Cossignani, L. Oxidative modifications of conjugated and unconjugated linoleic acid during heating. *Food Chem.* **2013**, *140*, 680–685. [CrossRef]

43. Williams, C.D.; Whitley, B.M.; Hoyo, C.; Grant, D.J.; Iraggi, J.D.; Newman, K.A.; Freedland, S.J. A high ratio of dietary n-6/n-3 polyunsaturated fatty acids is associated with increased risk of prostate cancer. *Nutr. Res.* **2011**, *31*, 1–8. [CrossRef]

44. FAO. *Fats and Fatty Acids in Human Nutrition: Report of an Expert Consultation*; Food and Agriculture Organization of the United Nations: Rome, Italy, 2010.

45. Keenan, D.F.; Resconi, V.C.; Smyth, T.J.; Botinestean, C.; Lefranc, C.; Kerry, J.P.; Hamill, R.M. The effect of partial-fat substitutions with encapsulated and unencapsulated fish oils on the technological and eating quality of beef burgers over storage. *Meat Sci.* **2015**, *107*, 75–85. [CrossRef]

46. Morales-Irigoyen, E.E.; Severiano-Pérez, P.; Rodriguez-Huezo, M.E.; Totosaus, A. Textural, physicochemical and sensory properties compensation of fat replacing in pork liver pate incorporating emulsified canola oil. *Food Sci. Technol. Int.* **2012**, *18*, 413–421. [CrossRef]

47. Rodríguez-Carpena, J.G.; Morcuende, D.; Estévez, M. Avocado, sunflower and olive oils as replacers of pork back-fat in burger patties: Effect on lipid composition, oxidative stability and quality traits. *Meat Sci.* **2012**, *90*, 106–115. [CrossRef]

48. Paneras, E.D.; Bloukas, J.G. Vegetable-oils replace pork backfat for low-fat frankfurters. *J. Food Sci.* **1994**, *59*, 725–728. [CrossRef]

49. Paneras, E.D.; Bloukas, J.G.; Filis, D.G. Production of low-fat frankfurters with vegetable oils following the dietary guidelines for fatty acids. *J. Muscle Foods* **1998**, *9*, 111–126. [CrossRef]

50. Bloukas, J.G.; Paneras, E.D. Substituting olive oil for pork backfat affects quality of low-fat frankfurters. *J. Food Sci.* **1993**, *58*, 705–709. [CrossRef]

Article

Impact of Culinary Procedures on Nutritional and Technological Properties of Reduced-Fat Longanizas Formulated with Chia (*Salvia hispanica* L.) or Oat (*Avena sativa* L.) Emulsion Gel

Tatiana Pintado, Claudia Ruiz-Capillas, Francisco Jiménez-Colmenero and Ana M. Herrero *

Institute of Food Science, Technology and Nutrition (ICTAN-CSIC), José Antonio Novais 10, 28040 Madrid, Spain; tatianap@ictan.csic.es (T.P.); claudia@ictan.csic.es (C.R.-C.); fjimenez@ictan.csic.es (F.J.-C.)
* Correspondence: ana.herrero@ictan.csic.es

Received: 14 November 2020; Accepted: 9 December 2020; Published: 11 December 2020

Abstract: This paper evaluates how grilling, a traditional culinary procedure for fresh meat products, affects the composition and technological properties of healthy longanizas formulated with chia (*Salvia hispanica* L.) (C-RF) and oat (*Avena sativa* L.) (O-RF) emulsion gels (EGs) as animal fat replacers. The use of EGs, regardless of whether they contain chia or oat, improved longaniza performance during cooking as they lost less ($p < 0.05$) water and fat. The composition of cooked sausages was affected by their formulation, particularly those with chia EG (C-RF) which featured the highest polyunsaturated fatty acid content, mainly due to the higher level of α-linolenic fatty acid (1.09 g/100 g of product). Chia and oat EGs in C-RF and O-RF allow longanizas to be labeled with nutritional and health claims under European law. In general, this culinary procedure increases ($p < 0.05$) the lightness, lipid oxidation and texture parameters of all samples.

Keywords: fresh sausages; longanizas; grilling; chia and oat EGs; nutritional and health claims; technological properties

---

1. Introduction

Plant-based ingredients are used to enhance health-promoting bioactive components in meat products [1,2]. Oat bran (*Avena sativa* L.) is an example that has been widely used in the preparation of a number of meat products, mainly providing them with soluble fiber (β-glucans), minerals (Mg, Fe, Cu, etc.), vitamins and phenolic compounds [3–5]. Chia (*Salvia hispanica* L.) is a plant-based ingredient that is gaining in popularity due to its interesting nutritional properties deriving from higher α-linolenic acid levels, insoluble dietary content and minerals (Ca, Fe, Mg, Mn, etc.) and vitamins [6]. Hence, chia has been added to different meat products to provide them with various bioactive compounds [7–13] and attractive technological properties (water and fat binding capacity, emulsifying and gelling properties, etc.) [14,15]. Due to their emulsifying and gelification potential, both chia and oat have been used in oil structuring processes to obtain new lipid materials which, thanks to their characteristics, can be used as animal fat substitutes. With this aim in mind, it is worth noting the use of EGs to reformulate fresh and cooked meat products due to their technological properties (texture, color, etc.) and their capacity to deliver bioactive compounds [16,17]. Especially in certain fresh meat products such as longanizas, chia and oat EGs have been deemed as suitable animal fat replacers and vehicles of bioactive compounds, allowing them to make different nutritional and health claims [18,19] and also feature optimal sensorial, technological and microbiological properties [20]. However, marketing these fresh sausages as healthier alternatives based on their composition may not be entirely accurate given the variety of processes that must be carried out prior to consumption (storage, thawing, cooking, etc.) [21]. Longanizas, for example, like other fresh meat products, need to be cooked prior to consumption. Cooking methods (roasting, grilling, frying, etc.) and conditions (time, temperature, heating rate, etc.) have a noticeable impact on the balance of (healthy/unhealthy) compounds and energy value [21]. Furthermore, lipid oxidation or other technological variations occur as a consequence of high temperatures [21,22].

It therefore makes sense that in the case of fresh meat products designed and produced to be healthier, we need information on their real nutritional and technological properties after having been subjected to different cooking procedures. Hence, the first aim was to produce a healthier fresh meat product (longanizas) by adding chia or oat EGs as animal fat replacers and delivery systems of certain bioactive compounds to evaluate the impact of grilling (selected because it is one of the most common ways to cook this type of product) on the composition and technological properties of longanizas. The second aim of the study was to determine how nutritional and health claims were affected by this cooking process. Longanizas elaborated with only animal fat (pork back-fat) (normal and reduced fat content) were used as reference.

2. Materials and Methods

2.1. Oil-in-Water Emulsion Gels and Longanizas Preparation

Oil-in-water (O/W) emulsion gels (EGs) and longanizas (fresh sausages) were prepared as reported by Pintado et al. [20]. Briefly, two different EGs, one containing chia (*Salvia hispanica* L.) flour and the other oat (*Avena sativa* L.) bran, were used as animal fat replacers in the preparation of longanizas. These EGs were formulated with 20% olive oil, 58% water, 2% gelling agent based on alginate (0.73% sodium alginate, 0.73% $CaSO_4$ and 0.54% sodium pyrophosphate) and 20% chia flour or oat bran depending on the desired formulation [16,17].

Four different longanizas were prepared with the same quantity of pork meat (60 g/100 g product). Two were formulated with only animal fat (pork back-fat) and used as the reference (all-animal fat): one with normal fat content (29 g pork back-fat/100 g product) called P-NF and the other with reduced fat content (7.25 g pork back-fat/100 g product) called P-RF. Additionally, two different reduced-fat longanizas were formulated replacing 90% of pork back-fat with 27 g/100 g of chia or oat EGs, with the final products denominated C-RF and O-RF, respectively. All the samples contained 4% of commercial seasoning for fresh sausages.

2.2. Cooking Method for the Longanizas and Weight Loss

Longanizas were cooked on an electric grill. Four samples were used for each formulation. This cooking method was chosen because it is the one that is most used for this kind of product. Preliminary cooking trials were performed to establish the cooking conditions required to achieve a meat core temperature of 70 °C. Sausages were weighed and cooked for 2 min at 210 ± 4 °C on an electric grill (Princess classic multigrill type 2321, Tilburg, The Netherlands). All sausages were then cooled at room temperature (20–22 °C) for 30 min, dabbed with a paper towel to remove visible exudate and then weighed to calculate weight loss determined by weight difference (four determinations). Results were expressed as a percentage of the initial weight. Samples were stored under chilled conditions (3 ± 2 °C) until analysis.

2.3. Composition and Energy Content of Longanizas

2.3.1. Proximate Composition

Moisture and ash contents were determined in triplicate according to official methods [23]. Fat content was evaluated in triplicate following Bligh and Dyer [24]. Protein was measured in quadruplicate with a Nitrogen Determinator LECO FP-2000 (Leco Corporation, St Joseph, MI, USA). All determinations were performed on both raw and grilled samples.

2.3.2. Fatty Acid Profile

Fatty acid content was determined (in triplicate for each type of sample) for both raw and grilled samples by saponification and bimethylation using C13:0 (7015-U Supelco PUFA No.2 Animal Source; Sigma-Aldrich Co., St. Louis, MO, USA) as the internal patron. Fatty acid methyl ester (FAME) was analyzed as described in Pintado et al. [20]. Fatty acids were expressed as g of fatty acid/100 g of product.

2.3.3. Energy Content

Energy content was calculated based on 9 kcal/g for fat and 4 kcal/g for protein and carbohydrate [25].

2.4. Technological Properties

2.4.1. Color and pH Determination

Color (CIE-LAB tristimulus values, lightness, L*; redness, a*; and yellowness, b*) was determined for raw and grilled sausage cross-sections using a Konica Minolta CM-3500 D Chroma Meter (Konica Minolta Business Technologies, Tokyo, Japan). Ten determinations were carried out for each sample type.

pH values were determined in quadruplicate for each formulation of longaniza using an 827 Metrohm pH-meter (MetrohmAG, Herisau, Switzerland) at room temperature on homogenates (1:10 *w/v* sample/distilled water ratio) of raw and cooked samples.

2.4.2. Texture Analysis

Kramer shear force (KSF) was performed using a miniature Kramer (HDP/MK05) cell with a 5-bladed head to perform a shearing test. Kramer shear tests were carried out at room temperature on sections of 2 cm previously weighed. A 25 kg load cell was used. Force was exerted to a compression distance of 25 mm at 0.8 mm/s crosshead speed using a TA-XT.plus Texture Analyzer (Texture Technologies Corp. Scarsdale, NY, USA). KSF values were calculated as the maximum force per g of sample (N/g). Five measurements were taken on both raw and grilled sausages for each formulation.

2.4.3. Lipid Oxidation Stability

Lipid oxidation was evaluated as a function of change in thiobarbituric acid-reactive substances (TBARs) [8]. TBAR determinations for each sample and formulation (raw and cooked) were performed in triplicate.

2.5. Statistical Analysis

One-way analysis of variance (ANOVA) was performed to evaluate statistical significance ($p < 0.05$) attributable to sample formulations (expressed in figures and tables with different superscript letters a, b, c, etc.) and two-way ANOVA was performed to evaluate statistical significance ($p < 0.05$) attributable to formulations and the effect of cooking (expressed in tables with different superscript numbers 1, 2) using the SPSS program (v.24, IBM SPSS Inc.; Chicago, IL, USA). Formulation and cooking were assigned as fixed effects and replication as a random effect. Results are given in terms of mean values and standard error of the mean. Least square differences (LSD) were used to compare mean values and formulations, while Tukey's HSD test was used to identify significant differences ($p < 0.05$) between formulations and cooking procedures.

3. Results and Discussion

3.1. Cooking Method for the Longanizas and Weight Loss

Fresh meat products such as longanizas must be cooked and the electric grill is commonly used. Figure 1 shows the different appearances of raw and grilled samples.

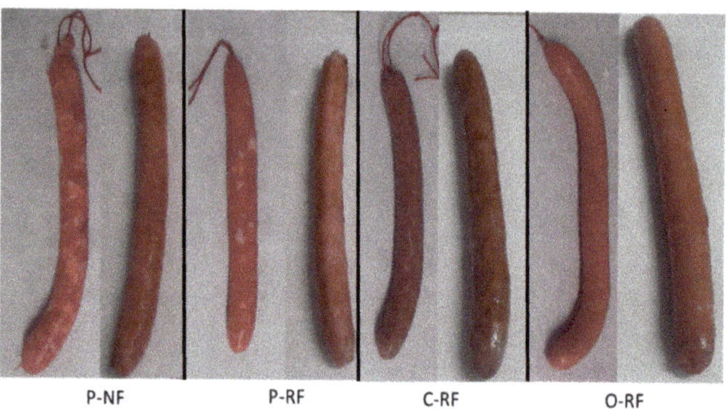

Figure 1. Appearance of longanizas: raw (left) and grilled cooked (right) (for each type of sample). Longanizas formulated with normal (P-NF) and reduced (P-RF) fat content (both with all-animal fat), and reduced-fat content sausages replacing 90% of pork back-fat with chia (C-RF) or oat (O-RF) emulsion gels (EGs).

This culinary procedure involves weight loss which furnishes information on the ability of the products to retain water and fat during processing that could alter the composition of sausages. Weight loss in longanizas cooked on an electric grill ranging from 6.71% to 24.73% was affected ($p < 0.05$) by formulation (Table 1) and could be considered between low and normal (15% and 40%) using comparable fresh meat products as the standard [26,27]. Samples with normal fat content (P-NF) exhibited greater ($p < 0.05$) weight loss (Table 1) than the reduced-fat samples, regardless of the type of fat. According to the literature, weight loss from culinary treatments tends to decrease as fat content decreases [28]. A comparison of reduced-fat sausages showed that those formulated with EGs (C-RF and O-RF), regardless of whether they contained chia or oat, lost less weight (up to 33% less than those made only with animal fat) (Table 1). It is important to note that these weight loss values are much lower than the ones observed for other reduced-fat fresh meat products where different structured lipid systems were used as animal fat replacers [26,27]. Moreover, our weight loss results are in accordance with water and fat binding properties associated with the thermal process (70 °C for 30 min in a water bath) in raw longanizas as samples reformulated with chia or oat EGs as fat replacers lost less weight than samples elaborated with all-animal fat [20].

Table 1. Weight losses and proximate composition (%) of grilled cooked longanizas.

Parameters	Samples *			
	P-NF	**P-RF**	**C-RF**	**O-RF**
Weight losses	24.73 ± 0.17 [c]	22.38 ± 0.27 [b]	6.71 ± 0.72 [a]	7.10 ± 1.13 [a]
Proximate composition				
Ash	3.89 ± 0.08 [c]	3.27 ± 0.06 [a]	3.49 ± 0.05 [b]	3.51 ± 0.03 [b]
Moisture	62.29 ± 1.04 [a]	70.93 ± 0.02 [c]	65.21 ± 0.56 [b]	66.40 ± 0.20 [b]
Protein	19.94 ± 0.33 [c]	17.29 ± 0.34 [b]	13.61 ± 1.27 [a]	14.97 ± 0.61 [a]
Fat	12.76 ± 0.53 [c]	6.92 ± 0.30 [a]	9.53 ± 0.75 [b]	7.92 ± 0.55 [a]

* For samples denominations, see Figure 1. Means ± standard deviation. Different superscript letters ([a,b,c]) in the same row (for each parameter) indicate significant differences ($p < 0.05$) between different formulations (P-NF, P-RF, C-RF and O-RF).

These results show that EGs are suitable fat replacers if the aim is to maintain a high level of fat and water in the final product after standard preparation on an electric grill. This may result in greater juiciness in the samples containing EGs as it has been shown that juiciness and cooking loss are negatively correlated, implying that low cooking loss results in greater juiciness [29]. Moreover, the lower cooking loss found in C-RF and O-RF could imply higher nutrient and bioactive compound stability [21]. This means that the use of EGs would result in greater retention of chia and oat bioactive compounds (α-linolenic acid, β-glucans, insoluble fiber, minerals, etc.) [16,17].

3.2. Composition and Energy Content of Longanizas

Longanizas are fresh meat products and must be cooked before eating and this could alter the concentration of some of their components [21]. Therefore, we have evaluated modifications in their composition resulting from grilling on an electric pan.

3.2.1. Proximate Composition

Slight formulation-related differences were observed in the composition of raw samples. Raw normal fat samples (P-NF) had the lowest ($p < 0.05$) moisture content (52%). However, a comparison among reduced-fat samples only (which ranged between 74 and 66%) showed that those prepared with EGs had the lowest ($p < 0.05$) values (approximately 66%). All raw longanizas had similar ($p > 0.05$) ash (~3%) and protein contents (13–14%). However, two different ($p < 0.05$) formulation-related fat levels were observed in raw samples: 30% in normal fat samples and approximately 13% in reduced-fat ones, regardless of whether they were made with animal fat or EG according to experimental design.

More differences were observed in the proximate composition of samples after grilling (Table 1). These significant differences could be mostly attributable to weight loss (Table 1) during the grilling process [21].

Regarding moisture content, cooked samples performed similarly to raw longanizas. In other words, reduced-fat samples exhibited higher ($p < 0.05$) values and, of these, samples with EG had the lowest ($p < 0.05$) moisture, regardless of whether they contained chia or oat (Table 1). These results coincide with observed weight losses (Table 1). Other authors have previously indicated that the moisture content of fresh grilled meat products such as patties, formulated in the same way as longanizas (by replacing animal fat with different types of emulsions), was lower than normal fat samples formulated with all-animal fat [26,27,30].

After cooking, normal fat samples (P-NF) exhibited the highest ($p < 0.05$) ash levels (related to mineral content) (Table 1). Among reduced-fat samples, those with EGs (C-RF and O-RF) had higher ($p < 0.05$) values than those with animal fat (P-RF) (Table 1). This could be because they lose less weight (Table 1) which implies lower mineral loss [31]. It must also be considered that these longanizas contain EGs made with chia and oat, ingredients with a high mineral content [16,17], which could account for the higher ash values observed in C-RF and O-RF (Table 1).

Cooked samples prepared with all-animal fat (P-NF and P-RF) had higher ($p < 0.05$) protein content than C-RF and O-RF, which had similar values regardless of whether they contained chia or oat (Table 1). This protein content in cooked longanizas could be related to the weight losses associated with each type of longaniza (Table 1) which, in turn, was possibly the result of moisture loss (drip and evaporation) and fat melting during cooking. In contrast to raw samples, three fat levels were observed in cooked samples (Table 1): normal fat samples (P-NF) with the highest ($p < 0.05$) fat content, and two significantly different reduced-fat levels, one higher level ($p < 0.05$) corresponding to C-RF and the other with similar values ($p < 0.05$) for P-RF and O-RF (Table 1).

Considering that the samples formulated with EGs (C-RF and O-RF) lost the same amount of weight (Table 1), it is worth noting that EG made with chia has a greater capacity to retain lipid content after grilling. A similar phenomenon has been described by other authors who observed a greater capacity to retain fat as a result of heat treatment in reduced-fat patties made with structured lipid materials as animal fat replacers when compared to samples made with animal fat only [26,27]. It is also interesting to note that, as a result of cooking, lipid content decreased by 60% in normal fat sausages compared to a 16–30% reduction in the case of P-RF, O-RF and C-RF.

3.2.2. Fatty Acid Profile

The fatty acid profile of longanizas, based on their saturated fatty acid (SFA), monounsaturated fatty acid (MUFA) and polyunsaturated fatty acid (PUFA) contents, is shown in Figure 2. Among raw samples, those with normal fat content exhibited the highest ($p < 0.05$) SFA and MUFA values (probably due to the higher lipid content). In reduced-fat samples, no significant differences were observed in MUFA content but SFA levels were higher in samples with all-animal fat (P-RF). Raw longanizas prepared with chia EG (C-RF) had similar ($p > 0.05$) PUFA content to samples with normal fat content (Figure 2), despite their lower ($p < 0.05$) lipid content (Table 1).

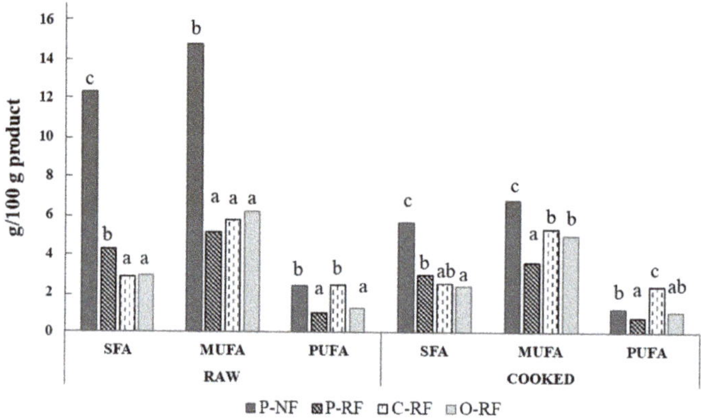

Figure 2. Saturated fatty acid (SFA), monounsaturated fatty acid (MUFA) and polyunsaturated fatty acid (PUFA) (g/100 g of product) of raw and grilled cooked longanizas. For samples denominations, see Figure 1. Different letters (a, b, c), for the same type of fatty acid (SFA, MUFA or PUFA) and treatment of samples (raw or cooked), indicate significant differences ($p < 0.05$) between different formulations (P-NF, P-RF, C-RF and O-RF).

Similarly, SFA and MUFA contents in cooked samples were higher ($p < 0.05$) in longanizas with normal fat content. However, among reduced-fat samples, those with chia or oat EGs exhibited higher MUFA values than longanizas with all-animal fat. Furthermore, samples with chia EG (C-RF) also showed the highest PUFA level: in most cases, double that of the others (Figure 2). It is worth highlighting their ALA content (1.09 g/100 g of product) even after cooking. Consequently, samples prepared with chia EG would be a good choice based on the daily dietary intake reference for ALA [32].

Use of these EGs in meat product development could be an interesting strategy for obtaining products with a healthy lipid profile, not only due to their vegetable oil content (such as olive oil) and other ingredients with high levels of healthy oil compounds (such as chia seed), but also due to their high capacity to retain that lipid content during cooking (Figure 2), a property not observed in all-animal fat samples (Table 1).

Evaluation of lipid composition is crucial in cooked products because meat fatty acids melt between about 25 and 50 °C, but SFAs melt at higher and PUFAs at lower temperatures. Moreover, changes in fatty acid concentration (mainly decreased PUFAs) can occur during cooking due to their low oxidative stability [21]. As a consequence of the culinary process applied, both SFA and MUFA content decreased in samples prepared with all-animal fat, ~54% in N-PF and ~30% in R-PF, while the decrease in PUFA was lower (49% and 22%, respectively, in these samples). This behavior was similar in longanizas made with oat EG, where SFA and MUFA contents decreased by approximately 19% for both, while PUFAs decreased by 16%. However, longanizas made with chia EG exhibited different reduction values for SFA, MUFA and PUFA (approximately 12%, 8% and 3%, respectively).

3.2.3. Energy Value

Normal fat samples exhibited the highest energy values in both raw and cooked samples. However, as a consequence of culinary treatment, the energy value in these samples (P-NF) decreased by about 40%, from 325 to 195 Kcal/100 g of product. This is due to the decreased fat content as a consequence of cooking. All raw reduced-fat samples (R-RF, C-RF and O-RF) exhibited energy values between 140 and 160 Kcal/100 g product which changed very little after grilling (approximately 130–140 Kcal/100 g of product).

3.2.4. Overall Nutritional Value: Nutrition and Health Claims

The initial contact that consumers have with food products is typically through their label. Hence, labels are important in creating consumer expectations. Products whose labels feature nutritional and health claims could improve consumer perception. These claims are made based on product composition at the time of purchase. Therefore, it is important to note that labels on raw reduced-fat longanizas, mainly those made with EGs, are allowed nutritional and health claims (Figure 3) under European legislation [18,19]. However, Article 10 of Regulation (EC) No 1924/2006, which lays down specific conditions for the permitted use of authorized health claims, considers it necessary and important to communicate the way the food is consumed, for example, after being cooked. In other words, consumers must be fully informed with regard to health claims made. Considering that cooking has an impact on the composition of longanizas, it stands to reason that analyses should be performed to ensure that these nutritional and health claims also apply to the cooked product. That is precisely why we have analyzed the impact that grilling has on the nutritional and health claims made with respect to raw longanizas. It is worth noting that this culinary procedure (grilling) had little or no effect on nutritional and health claims (Figure 3). Differences between raw and cooked products were only found in some samples with regard to "reduced fat" and "energy reduced" nutritional claims (Figure 3). Specifically, samples containing chia EG (C-RF) could not be labeled as "reduced fat" or "energy reduced" products after grilling on an electric pan, possibly due to the lower fat loss. Nevertheless, this could be a positive characteristic insofar as these samples maintain high ALA levels which benefit consumers [33].

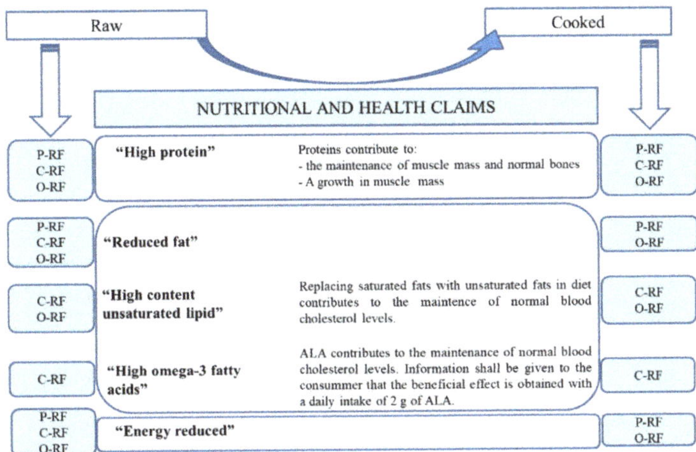

Figure 3. Nutritional and health claims of raw (left) and grilled cooked (right) longanizas. For samples denominations, see Figure 1; ALA: α-linolenic fatty acid.

3.3. Technological Properties

3.3.1. pH

pH values were between 6.09 and 6.39 and 6.18 and 6.44 for raw and cooked sausages, respectively (Table 2). Significant differences in pH values were observed resulting from formulation and culinary practices. Formulation-based differences were observed in samples reformulated with chia EG (C-RF) which showed the highest values for both raw and cooked sausages, while the lowest values corresponded to reduced-fat sausages with all-animal fat (P-RF) (Table 2). In most cases, cooking resulted in higher ($p < 0.05$) pH values. Similar behavior was observed in grilled meat products [34].

Table 2. pH values, color parameters ((L*) lightness, (a*) redness and (b*) yellowness)) and thiobarbituric acid-reactive substances values (TBARs) (expressed as mg MDA/kg sample) of raw and grilled cooked longanizas.

Parameters	Samples **	Raw	Grilled Cooked
	P-NF	6.09 ± 0.06 [b,1]	6.30 ± 0.02 [b,2]
pH	P-RF	6.01 ± 0.00 [a,1]	6.18 ± 0.01 [a,2]
	C-RF	6.39 ± 0.01 [d,1]	6.44 ± 0.02 [d,1]
	O-RF	6.22 ± 0.01 [c,1]	6.35 ± 0.02 [c,2]
Color parameters			
	P-NF	49.62 ± 4.82 [ab,1]	57.21 ± 1.21 [b,2]
L*	P-RF	46.22 ± 2.90 [a,1]	59.21 ± 0.98 [b,2]
	C-RF	48.52 ± 2.15 [ab,1]	54.68 ± 0.85 [a,2]
	O-RF	50.56 ± 1.33 [b,1]	58.06 ± 1.15 [b,2]
	P-NF	11.46 ± 1.03 [b,2]	9.47 ± 1.20 [b,1]
a*	P-RF	11.32 ± 1.00 [b,2]	9.78 ± 0.87 [b,1]
	C-RF	7.11 ± 0.99 [a,1]	6.30 ± 0.73 [a,1]
	O-RF	11.40 ± 1.20 [b,2]	8.72 ± 1.00 [b,1]
	P-NF	6.16 ± 0.80 [a,2]	4.49 ± 0.61 [a,1]
b*	P-RF	5.92 ± 0.43 [a,2]	4.49 ± 0.38 [a,1]
	C-RF	8.11 ± 0.94 [b,1]	7.41 ± 0.39 [b,1]
	O-RF	8.19 ± 0.70 [c,1]	7.43 ± 0.97 [b,1]
	P-NF	0.602 ± 0.08 [b,1]	0.801 ± 0.151 [b,1]
TBA(mg MDA/kg sample)	P-RF	0.123 ± 0.04 [a,1]	0.367 ± 0.093 [a,2]
	C-RF	0.392 ± 0.03 [a,1]	0.663 ± 0.140 [b,2]
	O-RF	0.348 ± 0.08 [a,1]	2.006 ± 0.241 [c,2]

** For samples denominations, see Figure 1. Means ± standard deviation. For each parameter, different superscript letters ([a,b,c,d]) in the same column indicate significant differences ($p < 0.05$) between different formulations (P-NF, P-RF, C-RF and O-RF). Different superscript numbers ([1,2]) in the same row indicate significant differences ($p < 0.05$) between formulations and treatments (raw and grilled) for each parameter.

3.3.2. Color

Color is important due to the impact it has on consumers' willingness to purchase meat products, with most finding bright red products more appealing. Instrumental color parameter (L*, a* and b*) values of longanizas were affected by the formulation and culinary procedure applied (Table 2). Regarding formulation, both raw and cooked sausages prepared with EG, regardless of whether they contained chia or oat, exhibited higher ($p < 0.05$) b* values. This could be related to the characteristic color of the olive oil used to make the EGs [16,17] which gave them a yellowish green hue due to pigments (mainly chlorophylls and carotenoids). Color likewise depends on the variety and ripeness of the fruit. Oat EGs did not have an impact on the typical red color of products of this sort because their a* values (for both raw and cooked) were significantly similar to those observed in samples made with animal fat (Table 2). However, both raw and cooked longanizas formulated with chia EGs (C-RF) had the lowest a* values which could be attributable to chia's dark color [8]. No significantly clear formulation-related differences were found for lightness (L*) (Table 2).

Lightness increased in all samples as a result of thermal treatment and redness decreased ($p < 0.05$) in all cases except for C-RF (Table 2). Parameter b* responded in two different ways: all-animal fat samples exhibited lower values after cooking, while the values of those prepared with EGs remained the same (Table 2).

3.3.3. Texture Analysis

Textural properties of grilled sausages can be useful in the development of a new product because those are the final textural attributes at the moment of consumption. The texture parameters of raw and cooked longanizas are shown in Figure 4. Raw longanizas exhibited similar ($p < 0.05$) KSF values but significant formulation-related differences were observed in cooked samples. KSF values increased for all samples as a result of cooking (Figure 4). Cooked samples made with all-animal fat (P-NF and P-RF) showed higher ($p < 0.05$) KSF values than those made with EGs (Figure 4). This could be related to the fact that normal and reduced-all-animal fat samples showed the highest ($p < 0.05$) weight loss from cooking (Table 1). It has been suggested that higher cooking losses may result in meat products that are more rigid and less easily broken in binding evaluations [35]. Similar textural behavior in terms of increased hardness from cooking was observed by other authors in fresh meat products such as patties reformulated using emulsions and bulking agents as animal fat replacers to improve their fat content [36,37].

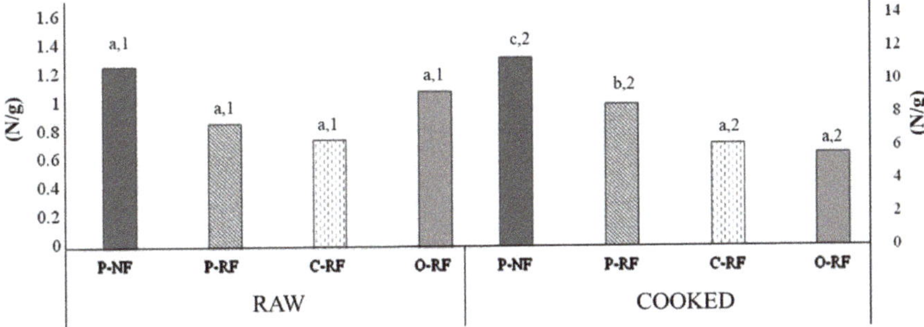

Figure 4. Kramer shear force (KSF, N/g) of raw and grilled cooked longanizas. For samples denominations, see Figure 1. Different letters (a, b, c), for the same treatment of samples (raw or cooked), indicate significant differences ($p < 0.05$) between different formulations (P-NF, P-RF, C-RF and O-RF). Different numbers (1,2) indicate significant differences ($p < 0.05$) between formulations and treatments for each type of sample.

3.3.4. Lipid Oxidation Stability

It is well known that thermal treatments such as grilling on an electric pan may accelerate lipid oxidation in meat products, promoting the development of off-flavors and the formation of potentially hazardous compounds. Moreover, high unsaturated fatty acid content, as found in the reformulated sausages used in this study (Figure 2), renders products more susceptible to lipid oxidation [38]. In this context, lipid oxidation levels of raw and grilled longanizas are shown in Table 2. Among raw samples, sausages with all-animal fat showed the highest ($p < 0.05$) lipid oxidation, while no significant differences were observed between reduced-fat samples. Cooking did not affect lipid oxidation in P-NF samples but in reduced-fat samples, it caused lipid oxidation products to accumulate (Table 2). However, the type of fat present in reduced-fat samples affected their oxidation stability and after cooking, those with oat EG (O-RF) showed the highest TBAR values (Table 2). This behavior may be related to the presence of beta-glucan in these samples as it has been shown that the oxidation process of β-glucan is fast at elevated temperatures (85 °C) [39]. Given their composition, lipid oxidation levels in the samples prepared with chia EG (C-RF) were unexpected in comparison with all-animal fat samples. In fact, C-RF even had higher PUFA levels than its all fat counterparts. This could be related to the higher levels of antioxidant compounds in chia seed [40]. Several studies have shown that oxidation caused by thermal treatment is very common in meat products that have been reformulated with lipid materials rich in unsaturated fatty acids [41–43].

4. Conclusions

Grilling appears to be a suitable cooking method to achieve healthier longanizas that have been formulated using chia or oat EGs as animal fat replacers. Especially after cooking, longanizas prepared with EGs maintained their composition, enabling them to make nutritional and health claims in accordance with European legislation. It should be noted that this reformulation strategy based on chia or oat EGs minimized weight loss during grilling compared with all-animal fat samples, thus improving yield. Furthermore, although the culinary process did modify some technological properties, for the most part, the behavior was similar for samples prepared with all-animal fat and those prepared with EGs.

Author Contributions: Conceptualization T.P., C.R.-C., and A.M.H.; data curation, T.P. and A.M.H.; formal analysis T.P.; funding acquisition, C.R.-C., F.J.-C. and A.M.H.; investigation, T.P., C.R.-C., F.J.-C. and A.M.H.; project administration, C.R.-C., F.J.-C. and A.M.H.; resources, C.R.-C., F.J.-C. and A.M.H.; writing—original draft, T.P., and A.M.H.; writing—review and editing, T.P., C.R.-C., F.J.-C. and A.M.H. All authors have read and agreed to the published version of the manuscript.

Funding: This research was funded by the Spanish Ministry of Science and Innovation (PID2019-107542RB-C21), by the CSIC Intramural projects (grant number 201470E073 and 202070E242), CYTED (grant number reference 119RT0568; HealthyMeat network) and the EIT Food Project 20206 (https://www.eitfood.eu/).

References

1. Zhang, W.; Xiao, S.; Samaraweera, H.; Lee, E.J.; Ahn, D.U. Improving functional value of meat products. *Meat Sci.* **2010**, *86*, 15–31. [CrossRef] [PubMed]
2. Hygreeva, D.; Pandey, M.C.; Radhakrishna, K. Potential applications of plant based derivatives as fat replacers, antioxidants and antimicrobials in fresh and processed meat products. *Meat Sci.* **2014**, *98*, 47–57. [CrossRef] [PubMed]
3. Yilmaz, I.; Daglioglu, O. The effect of replacing fat with oat bran on fatty acid composition and physicochemical properties of meatballs. *Meat Sci.* **2003**, *65*, 819–823. [CrossRef]
4. Santhi, D.; Kalaikannan, A. The effect of the addition of oat flour in low-fat chicken nuggets. *J. Nutr. Food Sci.* **2014**, *4*, 260.
5. Szpicer, A.; Onopiuk, A.; Poltorak, A.; Wierzbicka, A. The influence of oat beta-glucan content on the physicochemical and sensory properties of low-fat beef burgers. *CYTA J. Food* **2020**, *18*, 315–327. [CrossRef]
6. Munoz, L.A.; Cobos, A.; Diaz, O.; Aguilera, J.M. Chia seed (*Salvia hispanica*): An ancient grain and a new functional food. *Food Rev. Int.* **2013**, *29*, 394–408. [CrossRef]
7. Souza, A.H.P.; Gohara, A.K.; Rotta, E.M.; Chaves, M.A.; Silva, C.M.; Dias, L.F.; Matsushita, M. Effect of the addition of chia's by-product on the composition of fatty acids in hamburgers through chemometric methods. *J. Sci. Food Agric.* **2015**, *95*, 928–935. [CrossRef]
8. Pintado, T.; Herrero, A.M.; Jiménez-Colmenero, F.; Ruiz-Capillas, C. Strategies for incorporation of chia (*Salvia hispanica* L.) in frankfurters as a health-promoting ingredient. *Meat Sci.* **2016**, *114*, 75–84. [CrossRef]
9. Barros, J.C.; Munekata, P.E.S.; Pires, M.A.; Rodrigues, I.; Andaloussi, O.S.; Rodrigues, C.E.D.; Trindade, M.A. Omega-3-and fibre-enriched chicken nuggets by replacement of chicken skin with chia (*Salvia hispanica* L.) flour. *LWT* **2018**, *90*, 283–289. [CrossRef]
10. Ding, Y.; Lin, H.W.; Lin, Y.L.; Yang, D.J.; Yu, Y.S.; Chen, J.W.; Wang, S.Y.; Chen, Y.C. Nutritional composition in the chia seed and its processing properties on restructured ham-like products. *J. Food Drug Anal.* **2018**, *26*, 124–134. [CrossRef]

11. Antonini, E.; Torri, L.; Piochi, M.; Cabrino, G.; Meli, M.A.; De Bellis, R. Nutritional, antioxidant and sensory properties of functional beef burgers formulated with chia seeds and goji puree, before and after in vitro digestion. *Meat Sci.* **2020**, *161*, 108021. [CrossRef] [PubMed]

12. Camara, A.K.F.I.; Okuro, P.K.; da Cunha, R.L.; Herrero, A.M.; Ruiz-Capillas, C.; Pollonio, M.A.R. Chia (*Salvia hispanica* L.) mucilage as a new fat substitute in emulsified meat products: Technological, physicochemical, and rheological characterization. *LWT* **2020**, *125*, 109193. [CrossRef]

13. Pires, M.A.; Barros, J.C.; Rodrigues, I.; Munekata, P.E.S.; Trindade, M.A. Improving the lipid profile of bologna type sausages with Echium (*Echium plantagineum* L.) oil and chia (*Salvia hispanica* L) flour. *LWT* **2020**, *19*, 108907. [CrossRef]

14. Arendt, E.K.; Zannini, E. Oats. In *Cereal Grains for the Food and Beverages Industries*; Arendt, E.K., Zannini, E., Eds.; Woodhead Publishing: Philadelphia, PA, USA, 2013; pp. 243–282.

15. Valdivia-López, M.Á.; Tecante, A. Chia (*Salvia hispanica*): A review of native Mexican seed and its nutritional and functional properties. In *Advances in Food and Nutrition Research*; Jeyakumar, H., Ed.; Academic Press: Cambridge, MA, USA, 2015; Volume 75, pp. 53–75.

16. Pintado, T.; Ruiz-Capillas, C.; Jiménez-Colmenero, F.; Carmona, P.; Herrero, A.M. Oil-in-water emulsion gels stabilized with chia (*Salvia hispanica* L.) and cold gelling agents: Technological and infrared spectroscopic characterization. *Food Chem.* **2015**, *185*, 470–478. [CrossRef] [PubMed]

17. Pintado, T.; Herrero, A.M.; Jiménez-Colmenero, F.; Ruiz-Capillas, C. Emulsion gels as potential fat replacers delivering β-glucan and healthy lipid content for food applications. *J. Food Sci. Technol.* **2016**, *53*, 4336–4347. [CrossRef] [PubMed]

18. European Commission. Regulation (EC) No 1924/2006 of the European Parliament and of the Council of 20 December 2006 on nutrition and health claims made on foods. *Eur. Comm. Off. J. Eur. Union* **2006**, *404*, 3–18.

19. European Commission. Regulation (EU) No 432/2012 of the European Parliament and of the Council of 16 may 2012 establishing a list of permitted health claims made on foods other than those referring to the reduction of disease risk and to children's development and health. *Eur. Comm. Off. J. Eur. Union* **2012**, *136*, 1–40.

20. Pintado, T.; Herrero, A.M.; Jiménez-Colmenero, F.; Pasqualin Cavalheiro, C.; Ruiz-Capillas, C. Chia and oat emulsion gels as new animal fat replacers and healthy bioactive sources in fresh sausage formulation. *Meat Sci.* **2018**, *135*, 6–13. [CrossRef]

21. Jiménez-Colmenero, F.; Cofrades, S.; Herrero, A.M.; Ruiz-Capillas, C. Implications of domestic food practices for the presence of bioactive components in meats with special reference to meat-based functional foods. *Crit. Rev. Food Sci.* **2018**, *14*, 2334–2345. [CrossRef]

22. Sobral, M.M.C.; Cunha, S.C.; Faria, M.A.; Ferreira, I. Domestic cooking of muscle foods: Impact on composition of nutrients and contaminants. *Compr. Rev. Food Sci. Food Saf.* **2018**, *17*, 309–333. [CrossRef]

23. AOAC. *Official Methods of Analysis*, 18th ed.; Association of Official Analytical Chemists: Washington, DC, USA, 2005.

24. Bligh, E.G.; Dyer, W.J. A rapid method of total lipid extraction and purification. *Can. J. Biochem. Physiol.* **1959**, *37*, 911–917. [CrossRef] [PubMed]

25. EU Regulation. No 1169/2011 of the European Parliament and of the Council of 25 October 2011 on the provision of food information to consumers. *Eur. Comm. Off. J. Eur. Union* **2011**, *20*, 168–213.

26. López-López, I.; Cofrades, S.; Caneque, V.; Diaz, M.T.; López, O.; Jiménez-Colmenero, F. Effect of cooking on the chemical composition of low-salt, low-fat Wakame/olive oil added beef patties with special reference to fatty acid content. *Meat Sci.* **2011**, *89*, 27–34. [CrossRef] [PubMed]

27. Salcedo-Sandoval, L.; Cofrades, S.; Ruiz-Capillas, C.; Jiménez-Colmenero, F. Effect of cooking method on the fatty acid content of reduced-fat and PUFA-enriched pork patties formulated with a konjac-based oil bulking system. *Meat Sci.* **2014**, *4*, 795–803. [CrossRef] [PubMed]

28. Berry, B.W. Cooked color in high pH beef patties as related to fat content and cooking from the frozen or thawed state. *J. Food Sci.* **1998**, *63*, 797–800. [CrossRef]

29. Aaslyng, M.D.; Bejerholm, C.; Ertbjerg, P.; Bertram, H.C.; Andersen, H.J. Cooking loss and juiciness of pork in relation to raw meat quality and cooking procedure. *Food Qual. Prefer.* **2003**, *14*, 277–288. [CrossRef]

30. Alejandre, M.; Passarini, D.; Astiasaran, I.; Ansorena, D. The effect of low-fat beef patties formulated with a low-energy fat analogue enriched in long-chain polyunsaturated fatty acids on lipid oxidation and sensory attributes. *Meat Sci.* **2017**, *134*, 7–13. [CrossRef]

31. Maranesi, M.; Bochicchio, D.; Montellato, L.; Zaghini, A.; Pagliuca, G.; Badiani, A. Effect of microwave cooking or broiling on selected nutrient contents, fatty acid patterns and true retention values in separable lean from lamb rib-loins, with emphasis on conjugated linoleic acid. *Food Chem.* **2005**, *90*, 207–218. [CrossRef]
32. EFSA. Scientific opinion on dietary reference values for fats, including saturated fatty acids, polyunsaturated fatty acids, monounsaturated fatty acids, trans fatty acids, and cholesterol. *Eur. Food Saf. Agency J.* **2010**, *8*, 1461–1568.
33. WHO. *Diet, Nutrition and the Prevention of Chronic Diseases*; World Health Organization Technical Report Series; WHO: Geneve, Switzerland, 2003.
34. Serrano, A.; Librelotto, J.; Cofrades, S.; Sánchez-Muniz, F.J.; Jiménez-Colmenero, F. Composition and physicochemical beef steaks containing walnuts as characteristics of restructured affected by cooking method. *Meat Sci.* **2007**, *77*, 304–313. [CrossRef]
35. Shao, C.H.; Avens, J.S.; Schmidt, G.S.; Maga, J.A. Functional, sensory, and microbiological properties of restructured beef and emu steaks. *J. Food Sci.* **1999**, *64*, 1052–1054. [CrossRef]
36. Salcedo-Sandoval, L.; Cofrades, S.; Ruiz-Capillas, C.; Carballo, J.; Jiménez-Colmenero, F. Konjac-based oil bulking system for development of improved-lipid pork patties: Technological, microbiological and sensory assessment. *Meat Sci.* **2015**, *101*, 95–102. [CrossRef] [PubMed]
37. Freire, M.; Cofrades, S.; Serrano-Casas, V.; Pintado, T.; Jiménez, M.J.; Jiménez-Colmenero, F. Gelled double emulsions as delivery systems for hydroxytyrosol and n-3 fatty acids in healthy pork patties. *J. Food Sci. Technol.* **2017**, *54*, 3959–3968. [CrossRef] [PubMed]
38. Bernardi, D.M.; Bertol, T.M.; Pflanzer, S.B.; Sgarbieri, V.C.; Pollonio, M.A.R. ω-3 in meat products: Benefits and effects on lipid oxidative stability. *J. Sci. Food Agric.* **2016**, *96*, 2620–2634. [CrossRef] [PubMed]
39. Faure, A.M.; Nystroem, L. Metal binding capacity of oat beta-glucan affects its susceptibility to oxidation. *Abstr. Pap. Am. Chem. Soc.* **2014**, *247*, 31.
40. Scapin, G.; Schimdt, M.M.; Prestes, R.C.; Ferreira, S.; Silva, A.F.C.; Rosa, C.S.D. Effect of extract of chia seed (*Salvia hispanica*) as an antioxidant in fresh pork sausage. *Int. Food Res. J.* **2015**, *22*, 1195–1202.
41. Martínez, B.; Miranda, J.M.; Vázquez, B.I.; Fente, C.A.; Franco, C.M.; Rodríguez, J.L.; Cepeda, A. Development of a hamburger patty with healthier lipid formulation and study of its nutritional, sensory, and stability properties. *Food Bioprocess Technol.* **2012**, *5*, 200–208. [CrossRef]
42. Poyato, C.; Astiasaran, I.; Barriuso, B.; Ansorena, D. A new polyunsaturated gelled emulsion as replacer of pork back-fat in burger patties: Effect on lipid composition, oxidative stability and sensory acceptability. *LWT* **2015**, *62*, 1069–1075. [CrossRef]
43. Gómez-Estaca, J.; Pintado, T.; Jiménez-Colmenero, F.; Cofrades, S. The effect of household storage and cooking practices on quality attributes of pork burgers formulated with PUFA- and curcumin-loaded oleogels as healthy fat substitutes. *LWT* **2020**, *119*, 108909. [CrossRef]

Publisher's Note: MDPI stays neutral with regard to jurisdictional claims in published maps and institutional affiliations.

 *foods*

Review

Fruit and Agro-Industrial Waste Extracts as Potential Antimicrobials in Meat Products: A Brief Review

Leticia Aline Gonçalves [1,*], José M. Lorenzo [2,3] and Marco Antonio Trindade [1]

[1] Faculdade de Zootecnia e Engenharia de Alimentos, Universidade de São Paulo, Av. Duque de Caxias Norte, Pirassununga 13635-90, São Paulo, Brazil; trindadema@usp.br
[2] Área de Tecnología de los Alimentos, Facultad de Ciencias de Ourense, Universidad de Vigo, 32004 Ourense, Spain; jmlorenzo@ceteca.net
[3] Centro Tecnológico de la Carne de Galicia, Rúa Galicia N-4, Parque Tecnológico de Galicia, San Cibrao das Viñas, 32900 Ourense, Spain
* Correspondence: leticia.aline.goncalves@usp.br

Abstract: The use of antimicrobials in meat products is essential for maintaining microbiological stability. The reformulation by substituting synthetic additives for natural ones is an alternative to provide cleaner label products. Therefore, this work performed a literature search about extracts from fruits and agro-industrial waste with antimicrobial activity that can be applied in meat products. Jabuticaba waste extracts are excellent sources of anthocyanins with antimicrobial and pigmentation potential, capable of being applied in meat products such as fresh sausage, without compromising sensory attributes. Residue from grapes is rich in antimicrobial phytochemicals, mainly catechins, epicatechins, gallic acid and procyanidins. Extracts from different grape by-products and cultivars showed inhibition of *Staphylococcus aureus*, *Listeria monocytogenes*, *Pseudomonas aeruginosa*, *Escherichia coli* O157: H7 and other bacterial strains. Antimicrobial effects against *L. monocytogenes*, *Bacillus cereus*, *S. aureus* and *E. coli* O157: H7 were identified in *Opuntia* extracts. In addition, its application in hamburgers reduced ($p < 0.05$) aerobic mesophilic bacteria, *Enterobacteriaceae* and *Pseudomonas* sp. counts, and at a concentration of 2.5%, improved the microbiological stability of salami without causing sensory and texture changes. These data reinforce the possibility of substituting synthetic preservatives for natural versions, a growing trend that requires researching effective concentrations to maintain the sensory and technological properties.

Keywords: *Opuntia ficus-indica*; *Myrciaria cauliflora*; *Vitis* sp.; microbiological stability; sensory properties; cleaner label

Citation: Gonçalves, L.A.; Lorenzo, J.M.; Trindade, M.A. Fruit and Agro-Industrial Waste Extracts as Potential Antimicrobials in Meat Products: A Brief Review. *Foods* **2021**, *10*, 1469. https://doi.org/10.3390/foods10071469

Academic Editors: Claudia Ruiz-Capillas and Ana Herrero Herranz

Received: 13 May 2021
Accepted: 21 June 2021
Published: 25 June 2021

Publisher's Note: MDPI stays neutral with regard to jurisdictional claims in published maps and institutional affiliations.

1. Introduction

Since antiquity, even without being aware of the proliferation of microorganisms, when observing the high perishability of meat and the need for its immediate consumption, man began to use techniques of physical and chemical changes capable of delaying spoilage and improving the flavor of this and other food classes, which allowed the significant extension of the availability period of certain foods. One of the oldest forms of meat processing is the manufacture of by-products from the processing of meat pieces, which started around 1500 BC in the Mediterranean region, whose climate was favorable for the maturation of products, when several procedures that resulted in the reduction in water activity and consequently the prolongation of their shelf life, such as desiccation, drying, curing, smoking, salting and/or mixture of aromatic herbs, were also applied [1,2].

As they are nutritionally rich foods with a large amount of available water in their composition, meats become susceptible to contamination by pathogenic and spoiling microorganisms. In order to overcome this problem and offer safe meat products to consumers, it is necessary to adopt measures for their conservation, such as good manufacturing practices, use of low temperatures during storage, heat treatment and use of additives [3].

Processing has the purpose of extending the shelf life of meat, adding value to deboning by-products, which are generally not marketed in the fresh form, in addition to generating a wide variety of differentiated products in terms of color, flavor, aroma and texture [4]. Due to the low cost and easy preparation, a considerable part of the population developed the habit of regularly consuming meat derivatives such as sausages, bologna and hamburgers, contributing to a significant expansion of the meat products market [5,6].

The quality of meat derivatives is directly related to the origin of raw materials and ingredients and to the sanitary conditions of the manufacturing process. Products are classified according to the types of meat used, fat content, offal or edible by-products from slaughter, and may or may not be added with condiments and additives permitted by legislation [1]. In the meat production process, meat comminution increases the contact surface area, favoring microbial contamination and proliferation [7]. Therefore, the incorporation of preservatives in processing can considerably contribute to the maintenance of quality and safety characteristics during shelf life [8]. The use of these substances in meat products is regulated in Brazil by RDC No. 272/2019, which regulates the use of food additives for each class of meat derivative, their conditions of use and maximum limits [9].

For a long time, the food industry has incorporated various ingredients into formulations that do not have the function of nourishing, but rather have a technological purpose, while they can also make the food more attractive to consumers. These ingredients are called food additives and are classified according to their technological function [10]. The class of preservatives is one of them, the main purpose of which is to reduce the effect of spoilage caused by the multiplication of microorganisms or chemical reactions during the storage period [11,12]. Synthetic substances that have their use approved within an acceptable daily intake limit, such as nitrite and sodium nitrate, preservatives most commonly used in the production of meat derivatives, are also used, which in addition to their antimicrobial capacity, particularly for the control and prevention of the growth of anaerobic bacteria, especially *Clostridium botulinum*, also promote a protective effect against lipid oxidation and act in the development and fixation of the pink color and flavor characteristics of cured meat products [2,10,12].

The application of sodium nitrite in the production of cured meats allows obtaining differentiated products with regard to color and flavor, safe and stable during storage. However, a discussion that started around the 1970s showed the great risk to human health from the generation of a class of substances considered potentially carcinogenic, the nitrosamines, when high nitrite concentrations are exposed to high temperature conditions, as usually occurs in the manufacture of cured meat products, and since then, its use has been considered increasingly controversial [2,13].

Although they are substances that significantly contribute to the conservation of products and have their use officially regulated, there are indications of negative health implications associated with the excessive consumption of these and other synthetic additives, such as carcinogenic effects and generation of toxic and mutagenic compounds, and, consequently, the maximum acceptable limits of their use have been gradually changed or prohibited in several countries [14,15].

Diet is one of the important factors that affect the well-being and health of human beings, and today, there is great concern among consumers about the correlation between eating habits and health problems [16,17]. The reformulation of meat products through the substitution of ingredients, such as sodium nitrite, is an alternative to provide these products with a "cleaner label" in order to reduce the negative consumer perception about the excessive use of synthetic additives and their carcinogenic potential, decreasing the association between consumption of meat products and possible health problems [16,17]. For these reasons, many studies have been conducted in order to substitute synthetic antimicrobials for natural versions.

2. Phenolic Compounds

Phenolic compounds are phytochemicals with functions in pigmentation, astringency, protection against ultraviolet rays and antioxidant activity, being widely found in natural sources such as fruits, teas, spices, wine and honey [18]. These compounds have received much attention in recent decades due to evidence related to positive health effects, such as anti-inflammatory, antimicrobial, antithrombotic, vasodilatory and cardioprotective activity, contributing to the improvement in metabolic markers associated with diabetes, hypertension and obesity [19,20].

Phenolic molecules are structurally characterized by the presence of at least one aromatic ring containing one or more hydroxyl radicals, the main groups being phenolic acids, flavonoids and polyphenols, whose main source is fruits [18]. In addition, recent studies have reported antioxidant and antimicrobial effects of phenolic compounds, indicating that their chemical nature, especially the presence of hydroxyl groups in the molecule, may be associated with inhibitory mechanisms through interaction with the cytoplasmic membrane, cell wall and nucleic acids of bacteria, impairing vital functions such as protein synthesis and DNA transport or replication [21,22].

The generation of large amounts of waste from the processing of fruits and vegetables is one of the main challenges that the food industry has faced due to the need for large investment by companies to properly treat and dispose of this type of material in order to cause minimal negative impacts on the environment [14]. These agro-industrial waste products from fruits are rich in phenolic compounds and other bioactive substances that can add antioxidant and antimicrobial properties to foods and provide health benefits [23]. Thereby, the use of this raw material as a natural substitute for synthetic additives can be a great alternative, because in addition to providing compounds with functional properties, it reduces the environmental impact caused by the disposal of a significant part of the fruit, such as seeds and peels that are generally not used by the industry [24].

Thus, this review searched the scientific literature for reports of extracts obtained from fruits or their agro-industrial waste rich in antimicrobial bioactive compounds and that have potential applications in meat products, being able to maintain microbiological stability and safety during storage. More specifically, this review focused on the natural extracts obtained from jabuticaba, grape and prickly pear.

3. Fruit Extracts with Potential Application in Meat Products

3.1. Jabuticaba (Myrciaria cauliflora)

Jabuticaba (*Myrciaria cauliflora*), belonging to the *Myrtaceae* family, is a fruit tree native to Brazil, whose cultivation extends throughout the national territory, with greater productivity in the Southeastern region. From the nutritional point of view, jabuticaba varieties are considered excellent sources of dietary fibers, carbohydrates, vitamins and minerals such as iron, calcium and phosphorus, arousing great interest for its considerable amount of phenolic compounds with antioxidant and antimicrobial potential [25,26]. Among these compounds, anthocyanins and flavonoids are mainly concentrated in fruit peel, being the main components responsible for the development of its characteristic dark color. The anthocyanin content, of approximately 315 mg per 100 g of jabuticaba, is considered high compared to other fruits, demonstrating great potential as a substitute for synthetic dyes in several food classes, in addition to the benefits for the conservation of these products [8,27].

The most attractive source for obtaining these natural pigments from jabuticaba, rich in antioxidant and antimicrobial bioactive compounds, is the use of residues from jelly- and liquor-processing industries, since peels and seeds represent approximately 50% of the fruit and are in general discarded by the industry [28,29].

Baldin et al. [8] studied the application of microencapsulated jabuticaba extract in fresh sausage, evaluating the antimicrobial and antioxidant potential of this natural dye. Firstly, an in vitro experiment was carried out, which demonstrated an inhibitory effect on Gram-positive and Gram-negative bacteria, showing its antimicrobial potential. The minimum inhibitory concentration (MIC) results for the microencapsulated extract were

18.75 g/L (~2%) for both *Staphylococcus aureus* ATCC 25923 and *Escherichia coli* ATCC 25922. These results can be attributed to the high concentration of phenolic compounds (anthocyanins) in fruit peel, which are mainly responsible for the antimicrobial activity in this case [30].

When applying microencapsulated jabuticaba extract in fresh sausage, Baldin et al. [8] observed a reduction in the counts of mesophilic bacteria and of thermotolerant coliforms in treatments with 2 and 4% extract on the first and fifteenth days of cold storage. For *S. aureus*, treatments with 2 and 4% microencapsulated extract also showed lower counts when compared to the control (without addition of extract or dye) and with the treatment with added cochineal carmine dye. The addition of 4% extract caused the elimination of *S. aureus* on the last day of storage (15 days). In the total count of aerobic psychrotrophic microorganisms, a reduction of 1 log cycle was observed at the end of the fourth day of storage of treatments with additions of 2 and 4% of jabuticaba extract; however, from the eighth day, all treatments tested exceeded the limit of 10^7 CFU/g recommended by the International Commission on Microbiological Specifications for Foods (ICMSF) [31], indicating spoilage that can lead to sensory loss in the attributes of odor, color and taste. The count of lactic bacteria increased from 4 log CFU/g at time zero (beginning of storage) to 6 log CFU/g at the end of storage in all treatments, not exceeding the limit of 10^7 CFU/g established by ICMSF [31]. *Salmonella* sp. tested negative in 25 g for all treatments, being in accordance with Brazilian legislation [32].

Thus, the study recommended the addition of 2% of microencapsulated jabuticaba extract in fresh sausage, as it did not compromise the sensory attributes evaluated, except for the purplish color, which was slightly less accepted because it is not characteristic of the product. The aforementioned extract can be considered a good alternative for the production of cleaner label meat products as it satisfies the demand for the use of natural pigments with antimicrobial capacity and low cost, taking advantage of residues from the jabuticaba processing and with the appeal of health benefits.

3.2. Grape (Vitis sp.)

Grape (*Vitis* sp.) is one of the fruits most cultivated around the world, occupying an area of 7.5 million hectares of vineyards, with emphasis on the production of species *Vitis vinifera* in most countries [33,34]. In Brazil, the most commonly found cultivars are *Vitis labrusca*, *Vitis bourquina*, *Vitis vinifera* and several interspecific hybrids, occupying an area of 78 thousand hectares from the extreme south of the country to near the equator, showing characteristic poles of temperate, subtropical and tropical climates due to its expressive environmental diversity [35]. Brazilian grape production reached 1.5 million tons per year in 2018, with 50% destined for processing—wine making (42% table wines and 7% fine wines), juices (49%) and other derivatives (2%)—and the other half of the national volume marketed as grapes for fresh consumption [35].

The generation of waste from the processing of this high volume of grapes is significant, and may correspond to 30% of the fruit when used for the production of wines, for example, consisting of by-products such as pomace, peels and seeds [33,36]. This waste is considered a source of phenolic compounds with antioxidant and antimicrobial effects, mainly catechins, epicatechins, gallic acid and procyanidins [36–38].

Martin et al. [24] evaluated the in vitro antimicrobial capacity of ethanolic grape extracts from lyophilized seeds or agro-industrial waste. For the extract obtained from lyophilized pomace of "Pinot Noir" (*Vitis vinifera*) cultivar, the authors found MIC against Gram-positive *S. aureus* ATCC 25923 and *Listeria monocytogenes* ATCC 7644 pathogens of 6.25 and 12.5 g/L, respectively. "Petit Verdot" (*Vitis vinifera*) seeds ethanolic extract presented an MIC of 6.25 g/L for *L. monocytogenes* and 1.56 g/L for *S. aureus* (Table 1).

Table 1. Minimum inhibitory concentration in vitro of grape extracts on bacteria.

Strain	Gram	MIC	Grape Cultivar	Reference
		6.25 g/L	Pinot Noir (*V. vinifera*)	Martin et al. [24]
Staphylococcus aureus	+	1.56 g/L	Petit Verdot (*V. vinifera*)	Martin et al. [24]
		100.0 mL/L	Tempranillo (*V. vinifera*)	Adámez et al. [37]
Listeria innocua	+	100.0 mL/L	Tempranillo (*V. vinifera*)	Adámez et al. [37]
Listeria monocytogenes	+	12.5 g/L	Pinot Noir (*V. vinifera*)	Martin et al. [24]
		6.25 g/L	Petit Verdot (*V. vinifera*)	Martin et al. [24]

MIC: minimal inhibitory concentration.

Adámez et al. [37] estimated the in vitro antibacterial activity of aqueous grape seed extract (*Vitis vinifera* L.), Tempranillo cultivar, obtained after wine manufacture, and reported efficiency in inhibiting Gram-positive 976 *S. aureus subsp. aureus* and 910 *Listeria innocual* (Table 1). Similar results were obtained by Baydar et al. [34] in seeds from "Hasandede", "Emir" and "Kalecik" cultivars (all *Vitis vinifera* L. species), which demonstrated a relationship between increased extract concentration and reduced growth of 15 bacterial strains, including the pathogens *E. coli* O157: H7 KUEN 1461, *Aeromonas hydrophila* ATCC 7965 and *S. aureus* Cowan 1. Additionally, extracts at concentrations of 0.5 and 1% showed a bacteriostatic effect on *E. coli*, while concentrations of 2.5 and 5% provided bactericidal activity.

Despite the good in vitro results, data on the incorporation of this class of extract to guarantee microbiological stability in meat products were not found in thescientific literature. Carpes et al. [39] obtained lyophilized hydroethanolic (GPWL: grape pomace wine lyophilized) and microencapsulated (GPWM: grape pomace wine microencapsulated) extracts made with grape pomace from the processing of *Vitis labrusca* L. Bordeaux varieties and applied them to chicken pate in order to evaluate the effects of the addition of natural compounds on oxidative stability compared to negative control treatment (T1; no antioxidant added) and with the use of 300 ppm of the synthetic antioxidant sodium erythorbate (T2).

The study reported satisfactory results for the inhibition of lipid oxidation in treatments with 3000 ppm of GPWL (T3) and GPWM (T4) extracts, analyzed by the index of substances reactive to 2-thiobarbituric acid (TBARs) during 42 days of cold storage (4 °C). At the end of the storage period, all treatments showed a significant difference from each other ($p < 0.05$) in relation to the TBARS assay, with the best results being observed for T3, followed by T4, T2 and T1, respectively. All treatments, except for T1, had results below 3 mg of malondialdehyde/kg of sample, a value considered the limit for the meat product to be considered adequate and safe for consumption according to some authors [40,41]. Both GPWL and GPWM demonstrated an effective reduction in lipid oxidation when compared to treatment elaborated with commercial synthetic antioxidant, an activity that can be attributed mainly to the presence of phenolic compounds such as gallic, caffeic, vanillic, ferulic and coumaric acids and trans-resveratrol [39].

Thus, it appears that the extracts obtained from grape processing waste can be considered an interesting and innovative strategy for the incorporation of bioactive compounds in meat products with the substitution of synthetic preservatives for natural ones, since studies have shown their efficiency in inhibiting the growth of microorganisms related to outbreaks of foodborne diseases, such as *Listeria*, *E. coli* and *S. aureus*. However, studies on the in vivo influence on the sensory characteristics and antimicrobial action of meat products are necessary, since in general, studies have essentially evaluated the antioxidant activity.

3.3. Prickly Pear (*Opuntia ficus-indica*)

Opuntia ficus-indica, popularly known as prickly pear, is the fruit of a cactus species (*Cactaceae* family) native to tropical and subtropical regions of the Americas and currently also being cultivated in Europe, Africa and Australia, with approximately 300 known varieties [42]. The literature contains plenty of information about the chemical composition of its pulp, seeds and peel, as well as some properties of interest for the pharmaceutical and food industry, because it is a natural source of bioactive compounds.

Brazil has the largest *Opuntia ficus-indica* cultivation system in South America, with a planting area of 500,000 hectares located mainly in the Northeastern region and recently extended to other regions. Cultivation is performed in general by small producers and a large part of the production is destined for animal fodder, when it is called forage palm or cattle palm [43]. In the region of Valinhos, state of São Paulo, production is destined for the generation of fruit for fresh consumption, aimed at export to Europe and the domestic market [44]. In 2017, around 18.01 tons of the fruit were sold at "Companhia de Entrepostos e Armazéns Gerais de São Paulo" (CEAGESP), ranking 326th among products sold by the company [45].

In recent years, there has been a remarkable interest from the scientific community regarding the regular consumption of the genus *Opuntia* and its positive correlation with the treatment and prevention of chronic diseases related to oxidative stress [46,47]. Benefits such as reduction in triglycerides and total cholesterol in the bloodstream [48], antiulcerogenic activity [49], improved platelet aggregation [50] and reduced renal dysfunction [51,52] are some of the pieces of clinical and/or experimental evidence associated with the consumption of fig varieties. Other authors have found that extracts from the fruit and its peel and seeds have appreciable amounts of unsaturated fatty acids [53], with antioxidant activity [54,55], anticancer effects [56] and cardioprotective, antithrombotic, anti-inflammatory, antiarrhythmic, hypolipidemic and anti-hyperglycemic activities [57,58].

Seo et al. [59] identified the antimicrobial effects of *Opuntia ficus-indica* extract on two important pathogens related to foodborne diseases, *L. monocytogenes* and *E. coli* O157: H7, suggesting that the extract can be used as a natural preservative in food products. Zito et al. [60] detected the presence of eleven substances with antimicrobial activities in peels, seeds and pulps of the yellow (Surfarina) and red (Sanguigna) fruit varieties. Of these, major components were carvacrol, limonene, squalene and hexadecanoic acid, which in addition to their antimicrobial capacity are also antioxidants.

Parafati et al. [61] applied aqueous pulp extract of the purple and red *Opuntia ficus-indica* varieties in bovine hamburgers, testing the direct addition of the extract and the version encapsulated in sodium alginate. Microbiological analyses conducted after 8 days of cold storage (4 °C) showed the preservative effect in hamburgers with added prickly pear extract, which significantly reduced ($p < 0.05$) the count of mesophilic bacteria, *Enterobacteriaceae* and *Pseudomonas* sp., when compared to control sample with the addition of sterile distilled water. The authors concluded that the application of the extract, encapsulated or not, is an effective method for conservation of bovine hamburgers. However, studies are necessary to verify the influence of this component on the sensory and technological properties of products.

The antibacterial activity of the hydroethanolic extract obtained from the whole *Opuntia stricta* fruit, another species of the genus *Opuntia*, was quantified by determining the MIC and the minimum bactericidal concentration (MBC) by Kharrat et al. [62]. The results in Table 2 show that the extract from the red variety of *Opuntia stricta* showed high antibacterial activity, with an MIC and MBC less than or equal to sodium nitrite, which is the synthetic preservative commonly used in meat products, demonstrating that the extract can be as or more effective than sodium nitrite. This finding is mainly due to the presence of the pigment betalain and other phenolic compounds in the fruit.

When incorporating 2.5% of *Opuntia* extract in salami, replacing sodium nitrite preservative and the cochineal carmine dye, Kharrat et al. [61] obtained an improvement in the microbiological stability of products and in the water retention capacity without causing sensory and texture changes. The microorganisms surveyed were mesophiles, molds and yeasts, *S. aureus*, *Clostridium perfringens* and *Salmonella*, and all counts were within limits established by the legislation for both the control treatment and salami with added natural extract, concluding that this is a good alternative for maintaining shelf life during cold storage of this type of meat product.

Table 2. Minimum inhibitory concentration and minimum bactericidal concentration in vitro of *Opuntia* extract and sodium nitrite on bacteria.

Strain	Gram	MIC (mg/L)		MBC (mg/L)	
		OE	E250	OE	E250
Bacillus cereus	+	62.5	125.0	500.0	500.0
Staphylococcus aureus	+	62.5	62.5	250.0	500.0
Escherichia coli	−	125.0	250.0	500.0	>1000.0
Salmonella enteric	−	125.0	500.0	1000.0	1000.0

MIC: minimal inhibitory concentration; MBC: minimal bactericidal concentration; OE: *Opuntia* extract; E250: sodium nitrite. Adapted from Kharrat et al. [62].

Prickly pear and other *Opuntia* species, although showing an impressive profile of bioactive compounds, are not well valued in the country and in other parts of the world. Therefore, research on the antimicrobial capacity and the application of extracts from this fruit in meat products can be a good option for offering safe and cleaner label products.

4. Conclusions

The data shown in the present study reinforce the possibility of substituting, or at least reducing, the use of synthetic preservatives with natural versions, such as those from fruits and agro-industrial waste, such as peels and seeds. In addition to reducing the environmental impact through the reuse of residues that would initially be discarded, the use of preservatives from natural sources meets the demand of consumers who are increasingly concerned with the possible relationship between the indiscriminate use of synthetic additives and the incidence of health problems. Therefore, the search for plant extracts with antimicrobial activity is an area of study on the rise and there are several reports in the literature on natural alternatives, both for application in meat products and in other food classes, which, therefore, generates a challenge regarding the use of these ingredients in concentrations that are significantly effective against microbiological proliferation, while maintaining the sensory and technological properties of products without compromising their safety.

Author Contributions: Conceptualization, L.A.G., J.M.L. and M.A.T.; writing—original draft preparation, L.A.G.; writing—review and editing, M.A.T. and J.M.L.; visualization, J.M.L.; supervision, M.A.T. and J.M.L.; project administration, M.A.T. All authors have read and agreed to the published version of the manuscript.

Funding: This research received no external funding.

Acknowledgments: The authors are grateful to the São Paulo Research Foundation (FAPESP, Brazil) for financial support (grant number #2020/04544-9). This study was financed in part by the Coordenação de Aperfeiçoamento de Pessoal de Nível Superior–Brasil (CAPES)—Finance Code 001. Authors (M.A.T., J.M.L) are members of the HealthyMeat network, funded by CYTED (ref. 119RT0568).

Conflicts of Interest: The authors declare no conflict of interest.

References

1. Ordoñez, J.A.; Rodriguez, M.I.C.; Sanz, M.L.G.; Minguillón, G.D.G.F.; Perales, L.H.; Cortecero, M.D.S. *Tecnologia de Alimentos: Alimentos de Origem Animal*; Artmed: Porto Alegre, Brazil, 2005; p. 279. ISBN 9788536304311.
2. Sindelar, J.J.; Milkowski, A.L. Human safety controversies surrounding nitrate and nitrite in the diet. *Nitric Oxide* **2012**, *26*, 259–266. [CrossRef]
3. Terra, N.N. *Apontamentos de Tecnologia de Carnes*; Unisinos: São Leopoldo, Brazil, 1998; p. 216. ISBN 858558081X.
4. Pardi, M.C.; dos Santos, I.F.; de Souza, E.R.; Pardi, H.S. *Ciência, Higiene e Tecnologia Da Carne*; UFG e Universidade Federal Fluminense (Eduff): Goiania, Brazil, 2006; ISBN 9788572741880.

5. de Melo Filho, A.B.; Biscontini, T.M.B.; Andrade, S.A.C. Níveis de nitrito e nitrato em salsichas comercializadas na região metropolitana do recife. *Ciência Tecnol. Aliment.* **2004**, *24*, 390–392. [CrossRef]

6. Martins, L.L.; dos Santos, I.F.; Franco, R.M.; de Oliveira, L.A.T.; Bezz, J. Avaliação do perfil bacteriológico de salsicha tipo "Hot dog"comercializadas em embalagens a vácuo e a granel em supermercados dos municípios Rio de Janeiro e niterói, RJ/Brasil. *Rev. Inst. Adolfo Lutz* **2008**, *67*, 215–220.

7. Malavota, L.C.M.; Conte-Junior, C.A.; Macedo, B.T.; Lopes, M.M.; de Souza, V.G.; Stussi, J.S.P.; Pardi, H.S.; Mano, S.B. Análise micológica de linguiça de frango embalada em atmosfera modificada. *Rev. Bras. de Ciência Veterinária* **2006**, *13*. [CrossRef]

8. Baldin, J.C.; Michelin, E.C.; Polizer, Y.J.; Rodrigues, I.; de Godoy, S.H.S.; Fregonesi, R.P.; Pires, M.A.; Carvalho, L.T.; Fávaro-Trindade, C.S.; de Lima, C.G.; et al. Microencapsulated Jabuticaba (*Myrciaria Cauliflora*) extract added to fresh sausage as natural dye with antioxidant and antimicrobial activity. *Meat Sci.* **2016**, *118*, 15–21. [CrossRef]

9. Agência Nacional de Vigilância Sanitária. Resolução RDC N° 272, de 14 de Março de 2019. Estabelece os aditivos alimentares autorizados para uso em carnes e produtos cárneos. *Diário Of. União* **2019**, *52*, 194. Available online: https://www.in.gov.br/web/guest/materia/-/asset_publisher/Kujrw0TZC2Mb/content/id/67378977/do1-2019-03-18-resolucao-da-diretoria-colegiada-rdc-n-272-de-14-de-marco-de-2019-67378770 (accessed on 3 March 2020).

10. Lamas, A.; Miranda, J.; Vázquez, B.; Cepeda, A.; Franco, C. An evaluation of alternatives to nitrites and sulfites to inhibit the growth of salmonella enterica and listeria monocytogenes in meat products. *Foods* **2016**, *5*, 74. [CrossRef]

11. Hasiak, R.J.; Chaves, J.; Sebranek, J.; Kraft, A.A. Effect of sodium nitrite and sodium erythorbate on the chemical, sensory and microbiological properties of water-added Turkey ham. *Poult. Sci.* **1984**, *63*, 1364–1371. [CrossRef]

12. Clemente, F.; Marinelli, P.S.; Otoboni, A.M.M.B.; Tanaka, A.Y.; da Oliveira, A.S.; Nicolau, C.C.T. Verificação do teor de nitrito e nitrato em salsichas tipo hot dog em função dos métodos de cocção. *Rev. Anal.* **2014**, *1*, 72–78.

13. Cassens, R.G. Composition and safety of cured meats in the USA. *Food Chem.* **1997**, *59*, 561–566. [CrossRef]

14. Martínez, L.; Bastida, P.; Castillo, J.; Ros, G.; Nieto, G. Green alternatives to synthetic antioxidants, antimicrobials, nitrates, and nitrites in clean label Spanish chorizo. *Antioxidants* **2019**, *8*, 184. [CrossRef]

15. Verma, A.K.; Sharma, B.D.; Banerjee, R. Effect of sodium chloride replacement and apple pulp inclusion on the physico-chemical, textural and sensory properties of low fat chicken nuggets. *LWT Food Sci. Technol.* **2010**, *43*, 715–719. [CrossRef]

16. do Nascimento, R.; Campagnol, P.C.B.; Monteiro, E.S.; Pollonio, M.A.R. Substituição de cloreto de sódio por cloreto de potássio: Influência sobre as características físico-químicas e sensoriais de salsichas. *Aliment. Nutr. Braz. J. Food Nutr.* **2007**, *18*, 297–302.

17. Pires, M.A.; Munekata, P.E.S.; Villanueva, N.D.M.; Tonin, F.G.; Baldin, J.C.; Rocha, Y.J.P.; Carvalho, L.T.; Rodrigues, I.; Trindade, M.A. The antioxidant capacity of rosemary and green tea extracts to replace the carcinogenic antioxidant (BHA) in chicken burgers. *J. Food Qual.* **2017**, *2017*, 1–6. [CrossRef]

18. Lima, M.C.; Paiva de Sousa, C.; Fernandez-Prada, C.; Harel, J.; Dubreuil, J.D.; de Souza, E.L. A review of the current evidence of fruit phenolic compounds as potential antimicrobials against pathogenic bacteria. *Microb. Pathog.* **2019**, *130*, 259–270. [CrossRef]

19. Rice-Evans, C.; Miller, N.; Paganga, G. Antioxidant properties of phenolic compounds. *Trends Plant Sci.* **1997**, *2*, 152–159. [CrossRef]

20. Balasundram, N.; Sundram, K.; Samman, S. Phenolic compounds in plants and agri-industrial by-products: Antioxidant activity, occurrence, and potential uses. *Food Chem.* **2006**, *99*, 191–203. [CrossRef]

21. Xie, Y.; Yang, W.; Tang, F.; Chen, X.; Ren, L. Antibacterial activities of flavonoids: Structure-activity relationship and mechanism. *Curr. Med. Chem.* **2015**, *22*, 132–149. [CrossRef]

22. Sanhueza, L.; Melo, R.; Montero, R.; Maisey, K.; Mendoza, L.; Wilkens, M. Synergistic interactions between phenolic compounds identified in grape pomace extract with antibiotics of different classes against staphylococcus aureus and escherichia coli. *PLoS ONE* **2017**, *12*, e0172273. [CrossRef]

23. Brewer, M.S. Natural antioxidants: Sources, compounds, mechanisms of action, and potential applications. *Compr. Rev. Food Sci. Food Saf.* **2011**, *10*. [CrossRef]

24. Martin, J.G.P.; Porto, E.; Corrêa, C.B.; de Alencar, S.M.; da Gloria, E.M.; Cabral, I.S.R.; de Aquino, L.M. Antimicrobial potential and chemical composition of agro-industrial wastes. *J. Nat. Prod.* **2012**, *5*, 27–36.

25. Ascheri, D.P.R.; Ascheri, J.L.R.; Carvalho, C.W.P. De caracterização da farinha de bagaço de jabuticaba e propriedades funcionais dos extrusados. *Ciência Tecnol. Aliment.* **2006**, *26*, 897–905. [CrossRef]

26. Donadio, L.C. *Jaboticaba (Myrciaria Cauliflora (Vell) Berg)*; Funep: Jaboticabal, Brazil, 2000.

27. Bordignon-Luiz, M.T.; Gauche, C.; Gris, E.F.; Falcão, L.D. Colour stability of anthocyanins from isabel grapes (Vitis Labrusca L.) in model systems. *LWT Food Sci. Technol.* **2007**, *40*, 594–599. [CrossRef]

28. de Oliveira, A.L.; Brunini, M.A.; Salandini, C.A.R.; Bazzo, F.R. Caracterização tecnológica de jabuticabas "sabará" provenientes de diferentes regiões de cultivo. *Rev. Bras. Frutic.* **2003**, *25*, 397–400. [CrossRef]

29. Pereira, M.C.T.; Salomão, L.C.C.; Mota, W.F.; Vieira, G. Atributos físicos e químicos de frutos de oito clones de jabuticabeiras. *Rev. Bras. de Frutic.* **2000**, *22*, 16–21.

30. Caillet, S.; Côté, J.; Sylvain, J.F.; Lacroix, M. Antimicrobial effects of fractions from cranberry products on the growth of seven pathogenic bacteria. *Food Control* **2012**, *23*, 419–428. [CrossRef]

31. ICMSF (International Commission on Microbiological Specification for Foods). *Microorganisms in Foods—Sampling for Microbiological Analysis: Principles and Specific Applications*; University of Toronto Press: Toronto, ON, Canada, 1986; p. 193.

32. Brazil. Ministério da Saúde/Agência Nacional de Vigilância Sanitária/Diretoria Colegiada. Resolução RDC No 331, de 23 de Dezembro de 2019. Dispõe Sobre Os Padrões Microbiológicos de Alimentos e Sua Aplicação. Available online: https://www.in.gov.br/en/web/dou/-/resolucao-rdc-n-331-de-23-de-dezembro-de-2019-235332272 (accessed on 3 March 2020).

33. FAO. *Food and Agriculture Organization of the United Nations. Agribusiness Handbook: Grapes Wine*; FAO Investment Centre Division: Rome, Italy, 2009; Available online: http://www.fao.org/3/i7012e/i7012e.pdf (accessed on 16 April 2020).

34. Baydar, N.G.; Sagdic, O.; Ozkan, G.; Cetin, S. Determination of antibacterial effects and total phenolic contents of grape (Vitis Vinifera L.) seed extracts. *Int. J. Food Sci. Technol.* **2006**, *41*, 799–804. [CrossRef]

35. IBGE. Instituto Brasileiro de Geografia e Estatística. *Inteligência e Mercado de Uva e Vinho: A Viticultura No Brasil*. Available online: https://www.embrapa.br/cim-inteligencia-e-mercado-uva-e-vinho/a-viticultura-no-brasil (accessed on 20 April 2020).

36. Nardoia, M.; Ruiz-Capillas, C.; Casamassima, D.; Herrero, A.M.; Pintado, T.; Jiménez-Colmenero, F.; Chamorro, S.; Brenes, A. Effect of polyphenols dietary grape by-products on chicken patties. *Eur. Food Res. Technol.* **2018**, *244*, 367–377. [CrossRef]

37. Delgado Adámez, J.; Gamero Samino, E.; Valdés Sánchez, E.; González-Gómez, D. In Vitro Estimation of the Antibacterial Activity and Antioxidant Capacity of Aqueous Extracts from Grape-Seeds (*Vitis Vinifera* L.). *Food Control* **2012**, *24*, 136–141. [CrossRef]

38. Monagas, M.; Gómez-Cordovés, C.; Bartolomé, B.; Laureano, O.; Ricardo da Silva, J.M. Monomeric, oligomeric, and polymeric Flavan-3-Ol Composition of wines and grapes from Vitis Vinifera L. Cv. graciano, tempranillo, and cabernet sauvignon. *J. Agric. Food Chem.* **2003**, *51*, 6475–6481. [CrossRef]

39. Carpes, S.T.; Pereira, D.; de Moura, C.; dos Reis, A.S.; da Silva, L.D.; Oldoni, T.L.C.; Almeida, J.F.; Plata-Oviedo, M.V.S. Lyophilized and microencapsulated extracts of grape pomace from winemaking industry to prevent lipid oxidation in chicken pâté. *Braz. J. Food Technol.* **2020**, *23*, 1–13. [CrossRef]

40. Al-Kahtani, H.A.; Abu-Tarboush, H.M.; Bajaber, A.S.; Atia, M.; Abou-Arab, A.A.; El-Mojaddidi, M.A. Chemical changes after irradiation and post-irradiation storage in tilapia and Spanish mackerel. *J. Food Sci.* **1996**, *61*, 729–733. [CrossRef]

41. Selani, M.M.; Contreras-Castillo, C.J.; Shirahigue, L.D.; Gallo, C.R.; Plata-Oviedo, M.; Montes-Villanueva, N.D. Wine industry residues extracts as natural antioxidants in raw and cooked chicken meat during frozen storage. *Meat Sci.* **2011**, *88*, 397–403. [CrossRef] [PubMed]

42. Zakynthinos, G.; Varzakas, T. Lipid profile and antioxidant properties of selected pear cactus (*Opuntia Ficus-Indica*) ecotypes from southern greece. *Curr. Res. Nutr. Food Sci. J.* **2016**, *4*, 54–57. [CrossRef]

43. Taguchi, M.; Harinder Makkar, F.; Mounir Louhaichi, F.; Duffy, R.; Moretti, D.; Inglese, P.; Mondragon, C.; Nefzaoui, A.; Sáenz, C. Crop Ecology, Cultivation and Uses of Cactus Pear. Available online: http://www.fao.org/3/i7012e/i7012e.pdf (accessed on 16 April 2020).

44. Maiorano, J.A. Figo Da Índia—Ficha Técnica. 2016. Available online: https://www.todafruta.com.br/figo-da-india/ (accessed on 2 March 2020).

45. CEAGESP. Companhia de Entrepostos e Armazéns Gerais de São Paulo. Guia CEAGESP: Figo Da Índia. 2018. Available online: http://www.ceagesp.gov.br/guia-ceagesp/figo-da-india/ (accessed on 2 March 2020).

46. Osuna-Martínez, U.; Reyes-Esparza, J.; Rodríguez-Fragoso, L. Cactus (*Opuntia Ficus-Indica*): A review on its antioxidants properties and potential pharmacological use in chronic diseases. *Nat. Prod. Chem. Res.* **2014**, *2*, 1–8. [CrossRef]

47. Stintzing, F.C.; Herbach, K.M.; Mosshammer, M.R.; Carle, R.; Yi, W.; Sellappan, S.; Akoh, C.C.; Bunch, R.; Felker, P. Color, betalain pattern, and antioxidant properties of cactus pear (*Opuntia* spp.) clones. *J. Agric. Food Chem.* **2005**, *53*, 442–451. [CrossRef]

48. Palumbo, B.; Ethimiou, Y.; Stamatopoulos, J.; Oguogho, A.; Budinsky, A.; Palumbo, R.; Sinzinger, H. Prickly pear induces upregulation of liver LDL binding in familial heterozygous hypercholesterolemia. *Nucl. Med. Rev.* **2003**, *6*, 35–39.

49. Galati, E.M.; Mondello, M.R.; Giuffrida, D.; Dugo, G.; Miceli, N.; Pergolizzi, S.; Taviano, M.F. Chemical characterization and biological effects of Sicilian *Opuntia Ficus-Indica* (L.) Mill. Fruit juice: Antioxidant and antiulcerogenic activity. *J. Agric. Food Chem.* **2003**, *51*, 4903–4908. [CrossRef] [PubMed]

50. Wolfram, R.; Budinsky, A.; Efthimiou, Y.; Stomatopoulos, J.; Oguogho, A.; Sinzinger, H. Daily prickly pear consumption improves platelet function. *Prostaglandins Leukot. Essent. Fat. Acids* **2003**, *69*, 61–66. [CrossRef]

51. Park, E.-H.; Kahng, J.-H.; Lee, S.H.; Shin, K.-H. An anti-inflammatory principle from cactus. *Fitoterapia* **2001**, *72*, 288–290. [CrossRef]

52. Sreekanth, D.; Arunasree, M.K.; Roy, K.R.; Chandramohan Reddy, T.; Reddy, G.V.; Reddanna, P. Betanin a betacyanin pigment purified from fruits of *Opuntia Ficus-Indica* induces apoptosis in human chronic myeloid leukemia cell line-K562. *Phytomedicine* **2007**, *14*, 739–746. [CrossRef]

53. Chougui, N.; Tamendjari, A.; Hamidj, W.; Hallal, S.; Barras, A.; Richard, T.; Larbat, R. Oil composition and characterisation of phenolic compounds of *Opuntia Ficus-Indica* Seeds. *Food Chem.* **2013**, *139*, 796–803. [CrossRef]

54. Liu, H.G.; Liang, Q.Y.; Meng, H.L.; Huang, H.X. Hypoglycemic effect of extracts of cactus pear fruit polysaccharide in rats. *Zhong Yao Cai* **2010**, *33*, 240–242.

55. Matthäus, B.; Özcan, M.M. Habitat effects on yield, fatty acid composition and tocopherol contents of prickly pear (*Opuntia Ficus-Indica* L.) seed oils. *Sci. Hortic.* **2011**, *131*, 95–98. [CrossRef]

56. Abou-Elella, F.M.; Ali, R.F.M. Antioxidant and anticancer activities of different constituents extracted from Egyptian prickly pear cactus (*Opuntia Ficus-Indica*) Peel. *Biochem. Anal. Biochem.* **2014**, *3*, 1–9. [CrossRef]

57. Mobraten, K.; Haug, T.M.; Kleiveland, C.R.; Lea, T. Omega-3 and Omega-6 PUFAs induce the same GPR120-mediated signalling events, but with different kinetics and intensity in Caco-2 Cells. *Lipids Health Dis.* **2013**, *12*, 1–7. [CrossRef] [PubMed]

58. Berraaouan, A.; Ziyyat, A.; Mekhfi, H.; Legssyer, A.; Sindic, M.; Aziz, M.; Bnouham, M. Evaluation of antidiabetic properties of cactus pear seed oil in rats. *Pharm. Biol.* **2014**, *52*, 1286–1290. [CrossRef] [PubMed]

59. Seo, Y.H.; Han, C.H.; Lee, J.M.; Choi, S.M.; Moon, K.D. Effects of *Opuntia Ficus-Indica* extracts on inactivation of escherichia coli O157:H7 and listeria monocytogenes on fresh-cut apples. *J. Korean Soc. Food Sci. Nutr.* **2012**, *41*, 1009–1013. [CrossRef]

60. Zito, P.; Sajeva, M.; Bruno, M.; Rosselli, S.; Maggio, A.; Senatore, F. Essential Oils Composition of two Sicilian cultivars of *Opuntia Ficus-Indica* (L.) Mill. (*Cactaceae*) Fruits (Prickly Pear). *Nat. Prod. Res.* **2013**, *27*. [CrossRef] [PubMed]

61. Parafati, L.; Palmeri, R.; Trippa, D.; Restuccia, C.; Fallico, B. Quality Maintenance of Beef Burger Patties by Direct Addiction or Encapsulation of a Prickly Pear Fruit Extract. *Front. Microbiol.* **2019**, *10*. [CrossRef]

62. Kharrat, N.; Salem, H.; Mrabet, A.; Aloui, F.; Triki, S.; Fendri, A.; Gargouri, Y. Synergistic Effect of Polysaccharides, Betalain Pigment and Phenolic Compounds of Red Prickly Pear (*Opuntia Stricta*) in the Stabilization of Salami. *Int. J. Biol. Macromol.* **2018**, *111*, 561–568. [CrossRef]

Review

Development of Meat Products with Healthier Lipid Content: Vibrational Spectroscopy

Claudia Ruiz-Capillas and Ana M. Herrero *

Institute of Food Science, Technology and Nutrition (ICTAN-CSIC), José Antonio Novais 10, 28040 Madrid, Spain; claudia@ictan.csic.es
* Correspondence: ana.herrero@ictan.csic.es

Abstract: This review focuses on the importance of developing meat products with healthier lipid content and strategies such as the use of structured lipids to develop these enriched products. The review also conducts a critical analysis of the use of vibrational spectroscopy as a tool to further these developments. Meat and meat products are extensively recognized and consumed in the world. They are an important nutritional contribution in our diet. However, their consumption has also been associated with some negative consequences for health due to some of its components. There are new trends in the design of healthy meat products focusing mainly on improving their composition. From among the different strategies, improving lipid content is the one that has received the most attention. A novel development is the formation of lipid materials based on structured lipids such emulsion gels (EGs) or oil-bulking agents (OBAs) that offer attractive applications in the reformulation of health-enhanced meat products. A deeper interpretation is required of the complicated relationship between the structure of their components and their properties in order to obtain structured lipids and healthier meat products with improved lipid content and acceptable characteristics. To this end, vibrational spectroscopy techniques (Raman and infrared spectroscopy) have been demonstrated to be suitable in the elucidation of the structural characteristics of lipid materials based on structured lipids (EGs or OBAs) and the corresponding reformulated health-enhanced meat products into which these fat replacers have been incorporated. Future research on these structures and how they correlate to certain technological properties could help in selecting the best lipid material to achieve specific technological properties in healthier meat products with improved lipid content.

Keywords: healthier meat products; lipid content; structured lipids; emulsion gels; oil bulking agent; vibrational spectroscopy; Raman spectroscopy; infrared spectroscopy; technological properties

Citation: Ruiz-Capillas, C.; Herrero, A.M. Development of Meat Products with Healthier Lipid Content: Vibrational Spectroscopy. *Foods* **2021**, *10*, 341. https://doi.org/10.3390/foods10020341

Academic Editor: Thierry Astruc

Received: 19 December 2020
Accepted: 2 February 2021
Published: 5 February 2021

Publisher's Note: MDPI stays neutral with regard to jurisdictional claims in published maps and institutional affiliations.

1. Introduction

The global dynamics of the production and consumption of meat and meat products has evolved rapidly as the result of changing lifestyles and nutritional ideologies among part of the population. As a result, it is important to address several different aspects regarding the quality of meat and meat products, particularly nutritional (as relates to health), safety and sustainability aspects.

Regarding the nutritional aspects of meat and meat products, different strategies have been explored to optimize the composition of these products to make them healthier and to bring them in line with health recommendations and nutritional guidelines promoted by public bodies in response to new consumer demands.

Of the different strategies to improve meat and meat product composition, lipid content optimization has attracted the most attention owing to health recommendations [1]. This typically entails the partial or nearly complete substitution of fat with other healthier lipids by means of different technological procedures [1]. However, the most appropriate procedure in each case depends on the type of product, lipid material used, and the nature and magnitude of the proposed change. In the last few years, there has been growing attention in developing lipid materials to reduce or even completely replace fat in order to develop health-enhanced meat products. However, in most cases it is difficult to completely replace fat in food products. The physical and thermal properties of these products should be similar to those of animal fat but with fewer calories and an improved lipid profile [2,3]. A recent novel development is the formation and use of structured lipids such as emulsion gels (EGs) or oil-bulking agents (OBA) [4,5] that provide remarkable applications in the reformulation of health-enhanced meat products [2,6–10]. Although in some isolated cases, oil-structuring procedures applied to the reformulation of meat products as animal fat replacers could affect some of the properties of the final product, they have been successful in creating food that is viable in terms of technology, microbiology, sensorial properties and safety [2,3,6,7,9], with a shelf life similar to that of its standard meat counterpart. These animal fat replacers can improve the technological properties of meat products (water and fat content), and can also be employed as carriers of different nutrients (fiber, minerals, phenolic compounds, etc.), some with biological activity offering health benefits [6,7,9,11,12].

Successful development of structured lipid materials with healthier characteristics requires in-depth study of their nutritional, sensorial and technological characteristics, and a broader understanding of the complex relationship between the structure of their components and their properties. A new approach to the understanding of these interrelationships based on an analysis of the conformational modifications taking place within the components of lipid material and reformulated new meat products that affect their properties, merits consideration. In these connections, vibrational spectroscopic techniques such as Raman and infrared spectroscopy are direct and non-invasive techniques which offer an extensive range of possibilities in meat and meat products [6,7,13] and show great potential in supplementing structural information relating to fat-replacer production processes and the reformulation of health-enhanced meat products into which they are ultimately incorporated [4–7]. These techniques are also useful for quality control as it has been shown that their structure correlates with traditional ways of assessing meat quality (water retention capacity, texture, etc.) [13]

Based on the above, this review features a brief description of the novel strategies used to develop meat products with a healthier lipid content based mainly on the use of structured lipids (emulsion gels and oil bulking agents). It also proposes the employment of vibrational spectroscopy (infrared and Raman spectroscopy) to enhance the development of both structured lipids and the meat products which are included as animal fat substitutes. Moreover, the basic concepts of infrared and Raman spectroscopy are reviewed to aid in the understanding of how these spectroscopic techniques are applied.

2. Meat and Meat Products with a Healthier Lipid Content

Meat and meat products form part of the diet of many consumers in the world and are a significant source of an extensive range of nutrients essential for healthy development. These foods provide many nutrients such as high-value protein, fatty acids such as conjugated linoleic acid, minerals (iron, zinc and selenium) and B-complex vitamins. However, meat products also contain processing additives (sodium, nitrites, high fat content, etc.) with negative health implications. This has sparked the development of meat products more in line with health recommendations [14] using different procedures to enhance the composition of meat and especially meat products [1–3,15]. These procedures are mostly based on animal production (genetics and nutrition) and/or technological procedures focused on reducing or minimizing compounds with negative health implications and increasing compounds with positive health implications [1–3]. From among the technological strategies which aim to improve the global composition of meat products, special mention should be made to meat product reformulation processes which aim to remove, reduce, increase, add and/or replace certain components to develop healthier meat products [1–3,15].

From among the different reformulation processes, improved lipid content (in qualitative (lipid profile) and quantitative terms) is one of the most relevant processes in the development of healthier meat products [1–3,15] because there is an increasing indication of the connection between dietary fat intake and chronic disorders such as ischemic heart disease, some forms of cancer, obesity, and other ailments [14,16–18]. More specifically, some studies have shown a correlation between the intake of saturated fatty acids (SFAs) and an increase in total cholesterol levels in the blood. An increase in High-density lipoprotein (HDL) (ratio HDL/total cholesterol) is generally considered to be positive, whereas an increase in Low-density lipoprotein (LDL) is detrimental to health.

The consumption of SFA has been related to the development of cardiovascular disease, obesity, hypertension and some types of cancer [19]. The consumption of monounsaturated fatty acids (MUFAs), among which oleic acid stands out due to its high prevalence, has also been related to a reduction in LDL cholesterol and triglycerides in the blood. However, the role played by SFAs and MUFAs in human health is a controversial topic. Other studies question the health effects of SFAs and certain unsaturated fatty acids [20,21].

Despite this controversy, general consumption recommendations of total and unsaturated fatty acids can be made [14,22]. Briefly, dietary fat intake should preferably account for between 20–35% of total daily calories consumed [14,22]. According to dietary recommendations for the intake of specific fatty acids as a proportion of total diet, no more than 10% of calorie intake should be from SFAs, 6–10% from poly-unsaturated fatty acids (PUFAs) (n-6 and n-3, 6–11%), between 16–19% from MUFAs, and less than 1% from total fatty acids (TFAs) [14,22].

To meet these recommendations and improve the lipid content of meat products, technological strategies generally replace animal fat with different lipids, mainly from plant and marine sources, more in line with health recommendations (e.g., lower proportion of saturated fatty acids (SFA), higher in monounsaturated (MUFA) and polyunsaturated fatty acids (PUFA), especially from the omega-3 or n-3 family of fats). Efforts are also being made to improve PUFA content, the n-6/n-3 and PUFA/SFA ratios and, where possible, to reduce cholesterol and trans-fatty acids) [1]. This will predictably reduce the risk of developing the diseases discussed above.

Different procedures have been used to substitute animal fat with healthier lipids, ranging from the most conventional (direct addition, encapsulated, emulsified form, etc.) to the most novel recently developed techniques such as lipid structuring with EGs or the oil-bulking agents referred to above [1–3]. To that end, different vegetable oils (olive, soybean, etc.), marine oils (fish and algae), or mixtures of these have been employed to partially or completely substitute animal fat in various types of meat derivates (fresh, cooked, and dry cured). However, studies have shown that meat products reformulated in this way have different physicochemical characteristics which could impact negatively on the preferred quality parameters of the reformulated product [1–3]. Nevertheless, novel strategies such as lipid restructuring can be employed to enhance the quality of the reformulated meat products since these lipid materials and generate a solid fat which maintains solid-like properties. This could be a better way to develop meat products with a healthy lipid profile without negatively impacting their characteristics (Figure 1).

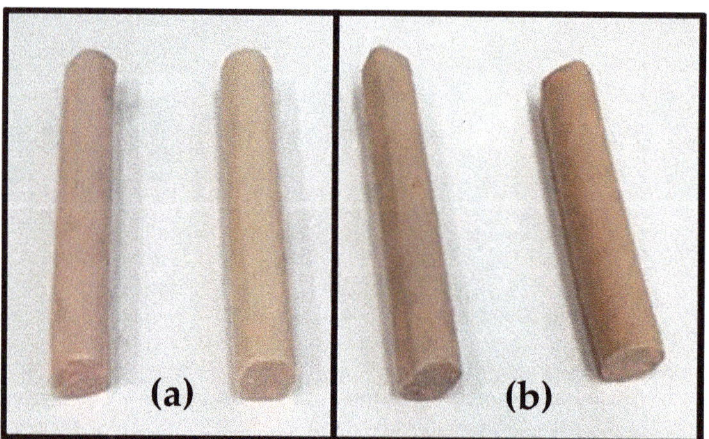

Figure 1. Example of healthy meat products with enhanced lipid content based on used structured lipids (as an animal fat replacer: (**a**) emulsion gels) and (**b**) oil bulking agent.

Lipid Materials Based on Structured Lipid: Emulsion Gels and Oil Bulking Agents

Lipid materials created using lipid structuring procedures have obtained a great deal of attention in the context of meat products. These lipid constituents can be employed as fat substitutes in meat products thanks to their solid-like properties simulating animal fat [1–3]. Their characteristics are of great interest for their application in meat products from a nutritional (healthy composition) and technological point of view. Of the various types of structured lipids EGs and OBAs are the most interesting due to their singular properties and possibilities as animal fat substitutes in meat products [1–3].

EGs are lipid materials in which emulsions and gels (hydrogels) concur. Formulation of these EGs fundamentally includes a lipid phase (olive, microalgal and chia oils, among others), an aqueous phase, an emulsifier (soy and whey protein, among others) and a gelling process [1–3]. Different compounds can be used in their formulation to give them particular technological characteristics, comprising specific bioactive compounds that provide nutritional benefits [2]. There are various procedures that can be used to create the gelling process such as heating, acidification, addition of divalent cations (Ca^{2+}), and enzymes with hydrocolloids [15,23]. Particular attention is being paid to cold gelling strategies employing binding agents such as alginates which form cold-set EGs [2,4].

Other interesting structured lipids are OBAs based on the dispersion of oil droplets in a continuous aqueous matrix-forming gel [2]. In these OBAs, liquid oil such as olive oil is usually enclosed in a hydrogel network structure. The formation of these OBAs includes mostly an early phase where the oil is distributed in the aqueous phase and lastly the gelation of the aqueous phase is induced by a gelling agent such as alginate. This procedure provides the OBAs with a solid structure, permitting it to be incorporate, for example, as an optimal fat replacer in meat products. Different gelling agents such as hydrocolloids (alginate) have been employed independently or in a mixture, causing diverse textures and structures in the final OBA [15].

Many studies have been conducted in recent years concentrating on the creation and characterization of EGs and their application in food products (yoghurt, cheese, sauces), particularly in the elaboration of cooked and fresh meat products (frankfurter type cooked sausages, fresh sausages, hamburgers, etc.) with a healthier lipid content [2,15,24,25]. However, OBAs have not been extensively employed in food science but their application as animal fat substitutes in healthier meat products is especially remarkable. These works have been focused mainly in the composition and technological characterization of EGS and OBAs. However, a further study of their structure and of the interactions between their diverse components, and their technological characterization, is required to gain greater insight into the different essential aspects necessary for their practical application as animal fat substitutes in meat products for the purpose of creating health-enhanced meat products. We likewise require a better interpretation of the changes in the structure of the main components found in meat products (proteins, lipids, etc.) that take place when structured lipids are incorporated as animal fat replacers, and how these affect the end product's technological properties. This knowledge will help to improve and optimize the development of health-enhanced meat products with specific technological characteristics. A useful approach could be to use vibrational spectroscopy (infrared and Raman spectroscopy) to analyze structural modifications in proteins and lipids in the formation of lipid materials and in the health-enhanced meat products in which these materials are incorporated as animal fat replacers.

To better understand the use of vibrational spectroscopy in the development of EGs and OBAs and in meat products with a healthier lipid content, it is necessary to adequately know the basic concepts of these techniques and the interpretation of the spectral results obtained with them.

3. Basic Concepts of Vibrational Spectroscopic Techniques: Infrared and Raman Spectroscopy

Vibrational spectroscopy, which involves infrared (IR) and Raman spectroscopies, is founded on the transitions concerning quantized vibrational energy states of molecules. In IR spectroscopy, the energy for these transitions is supplied by radiation in the IR regions (mid-IR and near-infrared) of the electromagnetic spectrum [26]. In Raman spectroscopy, samples are excited with monochromatic incident radiation that may be in the ultraviolet (UV), visible (VIS) or near-infrared (NIR) regions of the electromagnetic spectrum [26]. Complementary data on fundamental vibrational modes can be found from mid-IR and Raman spectra, as some vibrational motions are perceived mainly with IR radiation and others mostly by Raman scattering. In relation to protein structure, both vibrational spectroscopic techniques provide information about the secondary structure of proteins (α-helix, β-sheet, unordered), but Raman spectroscopy can also provide more detailed information about the tertiary structure of proteins [13,27–29]. Regarding structural changes in lipids, both provide relevant information (particularly in meat products with improved lipid content) and both have been used to provide information about the changes in their lipid structure. However, it should be noted that IR is faster and requires a smaller sample size than Raman spectroscopy [13,27–29].

IR and Raman spectroscopy provide many special benefits in food researches [13,27–29]. These techniques can be used to condensed-phase samples in several physical states, whether liquid or solid, clear or opaque. In many situations, insignificant or no sample pre-treatment is needed, and a spectrum can habitually be obtained quite fast [13,27–29]. There has been a remarkable increase in mid-infrared applications resulting from the progress of mid-infrared Fourier transform (FT-IR) spectrometers in combination with sampling methods comprising attenuated total reflection (ATR), for solids, semisolids and liquids, due to the benefits they offer [13,27–29]. Additionally, FT-IR microscopy has paved the way for novel uses of in situ microspectroscopic food mapping and imaging. In the last few years, hyperspectral imaging (HSI) has appeared as a hopeful analytical method for quality control and has involved a lot of attention in the non-destructive analysis of food products. Similarly, Fourier transform Raman spectroscopy (FT-Raman), using NIR excitation from a Nd: YAG laser at 1064 nm, can usually solve the drawback of fluorescence in foods [29]. Methods such as surface-enhanced Raman spectroscopy (SERS), confocal Raman microspectroscopy, and Raman imaging spectroscopy are established for their possibilities and special benefits in examining food components at very low amounts, and for in situ multi-component determination [26,30].

3.1. Analysis of Infrared and Raman Spectra Data

Infrared and Raman spectroscopy offer relevant data on the structure of the components of meat and meat products (proteins, lipids, water, etc.) non-invasively and in situ [26,30]. The reformulation processes used to develop meat products with an improved lipid content can modify Raman and infrared spectral bands due to structural changes in meat components (proteins, lipids and water). Basic information is therefore needed about how these changes are analyzed separately in proteins and lipids. For qualitative analysis, modifications can be visualized by comparing the spectrum in question with the characteristic bands of proteins, lipids or water. This requires an analysis of changes in the intensity, frequency, and half-widths of the Raman and infrared bands of chemical groups of proteins, lipids and water which are indicative of qualitative structural changes [13,27,29]. For quantitative analysis, curve-fitting of these bands is often used [13,27,29]. All this structural information is briefly described in the following sections.

3.1.1. Infrared Spectra

Mid-infrared (IR) spectroscopy is the most widely used spectroscopic technique in meat and meat products as it provides fundamental information about protein and lipid structure through well-resolved characteristic infrared bands. In recent years, mid-infrared Fourier transform (FT-IR) spectrometers with attenuated total reflection (ATR) stand out from among the different mid-infrared (IR) spectroscopy techniques employed to analyze the structure of meat and meat products.

The typical protein bands in the mid-infrared (IR) spectra are amide I (1700–1600 cm^{-1}), amide II (1560–1510 cm^{-1}) (Figure 2) and amide III (1300–1200 cm^{-1}) which provide information on protein secondary structure (α-helix, β-sheet, turn, unordered). Modifications in the intensity and/or frequency of these infrared bands is indicative of protein structural modifications [29,30]. The amide I band, high intensity (Figure 2) due to carbonyl stretching vibration with a slight influence from C-N stretching and N-H bending vibrations, is the one most frequently employed to analyze the secondary structure of proteins. Proteins with α-helical conformation show robust amide I bands between 1657 and 1650 cm^{-1}, whereas bands between 1640 and 1612 cm^{-1} are usually related with β-sheets.

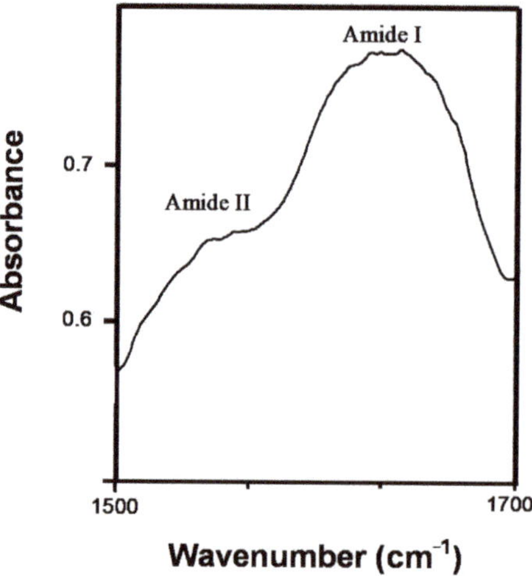

Figure 2. Typical FT-IR spectra from cooked sausage (type frankfurter) in the 1700–1500 cm^{-1}.

The mid-IR spectra of lipids encompass various bands in the 3000–1700 cm^{-1} region and overlapping bands in the 1500–700 cm^{-1} region. It is worth noting that in the elaboration of meat products with health-enhanced lipid content, infrared studies of the region between 3000–2800 cm^{-1} are characterized by two strong bands at 2925 and 2854 cm^{-1} (Figure 3) caused respectively from the asymmetric ($\nu_{as}CH_2$) and symmetric ($\nu_s CH_2$) stretching vibrations of the acyl CH_2 groups of lipids [31–33]. Modifications in lipid chain order/disorder causing from protein–lipid interactions can alter the half-bandwidth of these bands [34].

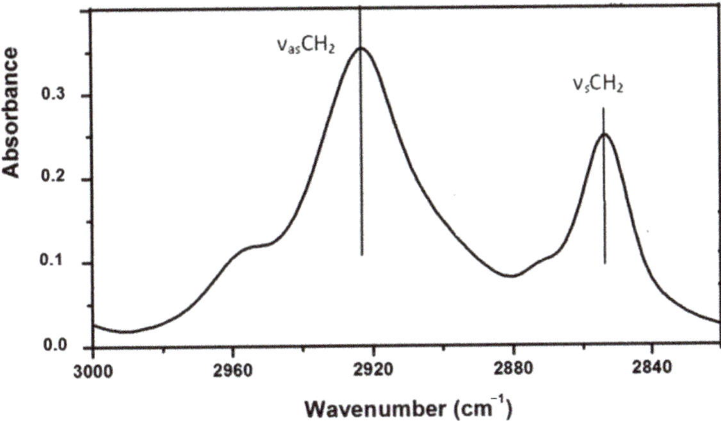

Figure 3. Representative FT-IR spectra in the 3000–2820 cm^{-1} region from cooked meat products (type frankfurter).

3.1.2. Raman Spectra

Raman spectroscopy offers data on protein structure, mostly by analysis of the amide I and III bands, which are associated with secondary structure. This spectroscopic technique also provides bands due to the environment of some side chains of proteins (aliphatic, tyrosine and tryptophan residues) and on the local conformations of disulfide bonds and methionine residues, all associated with a tertiary protein structure [27,29]. Fourier transform (FT) Raman spectrometers with 1064 nm excitation are commonly used to study the structure of these food products [13,27].

The principal Raman bands to establish the secondary structure of meat protein (α-helix, β-sheet, turn, unordered) are amide I (1650–1658 cm^{-1}), which involves mostly C=O stretching, and amide III vibrational modes (1225–1350 cm^{-1}) (Figure 4). Amide I, a strong band that involves mostly C=O stretching, is the most commonly used in the study of secondary protein structure. Most studies on the vibrational spectroscopy of proteins highlight the correlation between amide I band frequencies and protein secondary structure [27,29] proteins with great α-helical content, which display an amide I band centered about 1650–1658 cm^{-1} (Figure 4), while those with mainly β-sheet structures display the band between 1665 and 1680 cm^{-1}, and a great content of unordered structure is relate to proteins with an amide I band centered at 1660–1665 cm^{-1}. The spectral profile of the amide I band is employed in quantifying the secondary structure of proteins in terms of content of α-helix, β-sheet, turn and unordered using different methods mainly centered on frequency determinations at maximum intensity and half-bandwidth of the amide I band [27,29]. However, before this band can be analyzed, the water spectrum must first be correctly subtracted from the spectra. Additionally, other weaker bands can be noticed and analyzed in the Raman spectra. These correspond to the influence of peptide structure on the environment of some side chains such as those of aliphatic (δCHn), tyrosine (Tyr doublet) and tryptophan (Trp) residues (Figure 4), and on the local conformations of disulfide bonds and methionine residues associated to tertiary protein structure [35].

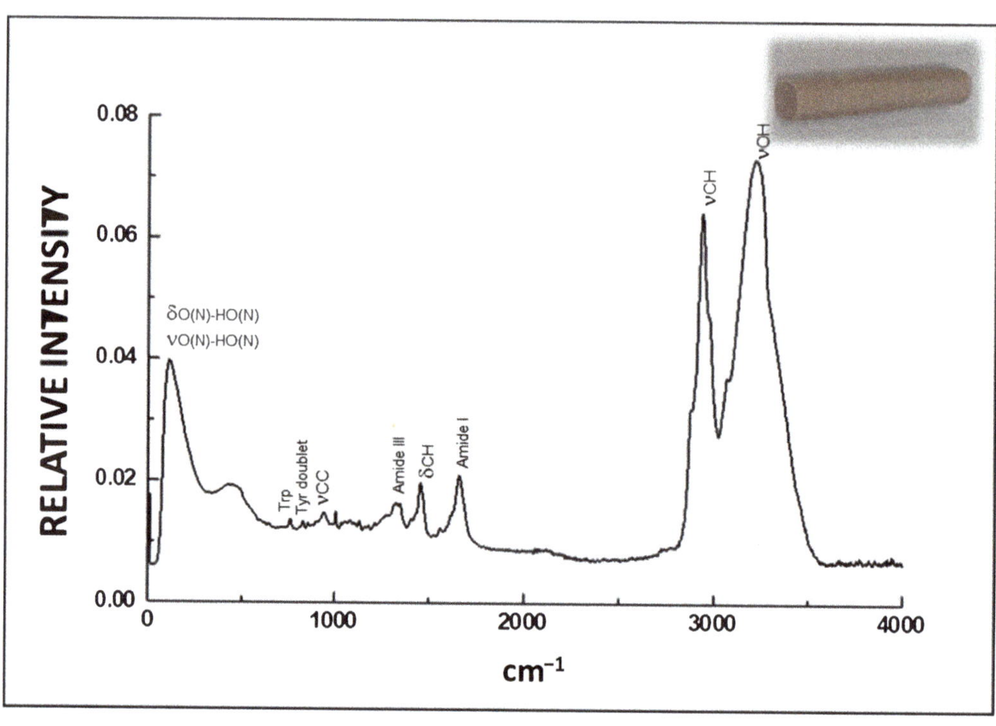

Figure 4. Typical Raman spectrum of a cooked sausage (type frankfurter) in the 0–4000 cm^{-1}.

There are also several Raman bands allocated to lipids near 1750, 1660, 1470, 1443, 1306, and 1269 cm^{-1} allocated to the C=O stretching modes, C=C stretching modes, CH_2 scissoring modes, CH_2 twisting modes, and CH in plane deformation modes of lipids [6,7]. The unsaturation level of fat-containing food products can be assessed precisely by analyzing the C=C stretching band (1660 cm^{-1}). Nevertheless, one of the most frequently used regions of the Raman spectra in the structural study of lipids in meat and meat products, especially those products with health-enhanced lipid content, is between 2800–3000 cm^{-1} (Figure 4) associated with changes in C-H stretching vibrations (νCH) [6,7]. A CH_3 symmetric stretching band near 2897 cm^{-1}, a CH_2 asymmetric stretching band near 2930 cm^{-1} and a CH_2 symmetric stretching motion near 2850 cm^{-1} are found in this region [29]. The symmetric and asymmetric vibrational modes of the CH_2 and CH3 groups can offer information about interactions between hydrocarbon chains. The peak height intensity relations $Iv_sCH_2/Iv_{as}CH_2$ (I2850/I2890) and $Iv_sCH_3/Iv_{as}CH_3$ (I2935/I2890) offer valuable indices to measure lipid packing consequences and establish relative order/disorder of the intermolecular lipid chain [36,37].

Raman spectra also show a broad water band (3100 and 3500 cm^{-1}) (Figure 4), associated with OH stretching motions [38,39], and a Raman band in the low-frequency range (below 600 cm^{-1}) including the bending (δ) and stretching (ν) vibrations of the O(N)-H..O(N) units, due to interactions of hydrogen bonded to water and protein molecules [38,39].

4. Application of Vibrational Spectroscopy in Meat Products with Healthier Lipid Content

Once this basic knowledge about vibrational spectroscopic techniques has been established, their application on the development of EGs and OBAs and healthier meat products in which these structured lipids are incorporated will be described separately.

4.1. Infrared and Raman Spectroscopy to Study Lipid Material Based on Structured Lipids

Many researchers have studied the possibilities of vibrational spectroscopy to establish fat content and the fatty acid composition of animal fat (adipose tissue from beef, lamb, pork and chicken, etc.) [40–43] Raman spectroscopy has been used to predict PUFA, MUFA and SFA content, and the degree of unsaturation (IV) in melted fat and adipose tissue of pork [41]. Raman spectral regions between 775 and 1800 cm^{-1}, and between 2600 and 3100 cm^{-1}, were selected for regression models since these contain bands related to lipids. Results showed that Raman spectroscopy is an interesting technique to measure the fatty acid composition of pork adipose tissue with the benefit that this technique is non-invasive and measurements can be completed online [41]. Vibrational spectroscopy has also been used extensively to achieve structural information about oils [31,32,43–45], which are the basis of lipid materials. Based on the useful results obtained from the analysis of animal fat and oils, many studies were later conducted on lipid materials, especially structured lipids.

FT-IR has been employed to study conformational modifications of oils emulsified with proteins (α-lactalbumin and β-lactoglobulin) where these proteins are adsorbed in the emulsion formation process [46,47]. This process induces changes in their secondary structure [46,47]. The results showed the creation of an intermolecular, antiparallel β-sheet upon adsorption due to protein self-aggregation. Studies to obtain details on protein secondary structures of olive oil-in-water emulsions stabilized with various protein systems based on caseinate or soy protein have been conducted using FT-IR [48,49]. The relationship between emulsion structure and its physical properties was also evaluated. Protein secondary structures changed to more orderly protein backbones, mainly involving the α-helical structure upon creation of the olive oil-in-water emulsion. These structural properties could be associated with the firmer textural characteristics found in soy and caseinate emulsions. A better interpretation of the potential relationship between the structural and textural characteristics of olive oil-in-water emulsions could help in selecting the best emulsion components to obtain specific textural characteristics. In this connection, FT-IR and FT-Raman play an important role. Structured lipids such as emulsion gels (EGs) are an interesting possibility to structuring edible oils. These are described as emulsions with a gel-like network structure and solid-like mechanical characteristics [2,3], making them a good alternative to animal fat in meat products with a healthy lipid profile. It is relevant to note the function of structure and lipid phase interactions in the stability of EGs and their technological properties [4,5]. IR, especially attenuated total reflectance (ATR)-FTIR spectroscopy, has proven valuable insofar as it is a fast, non-destructive analytical method capable of offering analytical and structural information on various EGs. Molecular structures of polyphenol–soy EGs have been studied using this spectroscopic technique owing to their potential as release systems for bioactive compounds (polyphenols, MUFAs, PUFAs). Studies on polyphenol–soy EGs have shown that phenolic compounds seem to become confined in the emulsion matrix network, due to its textural properties, exhibiting lower gel strength than EGs without polyphenols [50]. ATR-FTIR spectroscopy has also been used to examine the structural characteristics of chia EGs. Analyses of the 3000–2500 cm^{-1} region indicate that the order/disorder of the oil lipid chain, associated to lipid interactions and droplet size in the emulsion gels, could be important in controlling their textural properties [4]. Raman spectroscopy has also been used to examine the lipid structure of different chia EGs and spectroscopic results, mainly significant differences were found in the area 2800–3000 and I2854/I2900, and indicated differences depending on these EG compositions in the lipid structure and interactions in terms of lipid acyl chain mobility (order/disorder). These lipid structural properties of chia oil EGs linked significantly with a particular textural behavior [9].

Polysaccharides, a structured lipid, can be employed either individually or in mixture to form a diversity of gel structures which may be appropriate for immobilizing oil droplets working as oil-bulking agents. Raman spectroscopy has also been employed to determine lipid and polysaccharide interactions in different polysaccharide gels elaborated for use as oil-bulking agents [5]. Raman spectroscopic results show that these structured lipids were stabilized by hydrogen bonding between oil carbonyl groups and water and/or carbohydrate molecules. This structural behavior may clarify the differences in textural characteristics. Structural studies of lipid materials (EGs and OBA) contribute to a better interpretation of the correlation between their technological properties and the structural characteristics of their ingredients. Understanding these correlations ultimately helps in selecting the most suitable materials.

This information is of particular interest given that these lipid materials can help to improve or maintain the quality of the food to which they are added. Particularly important in this regard is their use as animal fat replacers, for instance in the development of health-enhanced meat products without compromising the properties of the final product.

4.2. Infrared and Raman Spectroscopy to Examine Meat Products with Healthier Lipid Content

Vibrational spectroscopic techniques, particularly FT-Raman and FT-IR, provide information about the secondary structure of proteins and the structure of lipids in meat and meat products [6,13,27,28]. These spectroscopic techniques have been used to build models that can predict sensory and technological characteristics, and also to evaluate structural modifications appearing in proteins during processing [13,28,29]. Studies evaluating the feasibility of FT-IR spectroscopy in determining structural properties have focused on health-enhanced meat products reformulated with lipid material such as olive oil-in-water emulsions stabilized with casein or soy protein to replace animal fat [51]. The results of these studies suggest that this spectroscopic technique offers relevant information about changes in protein secondary structure caused by changes in the amide I band profile. Moreover, changes in the 3000–2500 cm^{-1}, particularly the half-bandwidth in the 2922 cm^{-1} band, has suggested modifications in lipid chain order or lipid–protein interactions. These results underscore the relevance of the stabilizing system employed in the preparation of oil-in-water emulsions as it impacts both the structural and technological characteristics of the reformulated meat products.

Lipid materials based on structured lipids such as chia EGs have been applied as animal fat substitutes to develop frankfurters with a health-enhanced lipid profile. In this case, ATR-FTIR spectroscopy was used to study how replacing animal fat with chia EGs affects the structural characteristics of lipids [52]. Results from the 3000–2500 cm^{-1} region showed that frankfurter reformulated with chia emulsion gels implicate more lipid–protein interactions which correlated significantly with processing loss (mainly loss of water and fat) and the textural behavior of these samples [52]. ATR-FTIR spectroscopy was also used to study health-enhanced frankfurters reformulated with structured lipids based on polyphenol-EGs as animal fat replacers (Figure 5). The spectroscopic result of the acyl chain region (between 2950 and 2830 cm^{-1}) of the ATR-FTIR spectrum presented greater ($p < 0.05$) inter- and intramolecular lipid disorder irrespective of the presence of polyphenol in the EG [53].

FT-Raman spectroscopy was also employed to study the structural properties of health-enhanced frankfurters reformulated with structured lipids such as oil-bulking agents to replace fat [6]. Analysis of the amide I band furnished valuable information on secondary protein structures regarding enrichment of β-sheet structures in these frankfurters reformulated with oil-bulking agents [6]. The high content of β-sheet structures in these health-enhanced frankfurters indicate the creation of a denser network, resulting in increased hardness. Regarding lipid structure, analysis of the 2800–300 cm^{-1} regions and intensity ratios $Iv_sCH_2/Iv_{as}CH_2$ ($I2850/I2890$) and $Iv_sCH_3/Iv_{as}CH_3$ ($I2935/I2890$) revealed that the addition of oil-bulking agents as animal fat replacers in these health-enhanced frankfurters increased lipid acyl chain disorder and implicated more lipid–protein interactions [6].

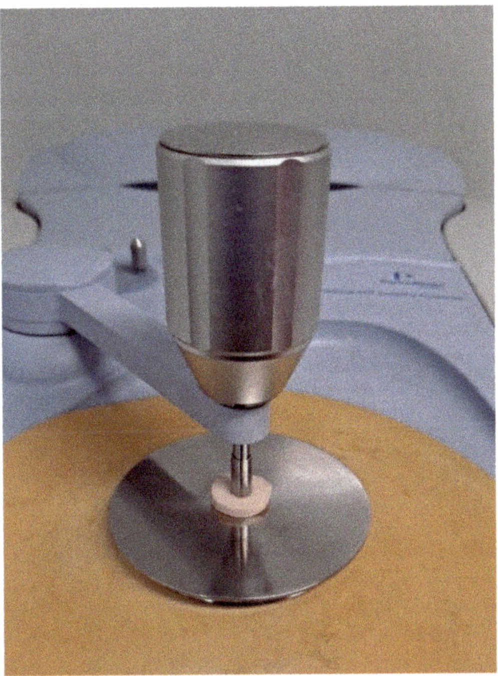

Figure 5. Analysis of frankfurters reformulated with structured lipids based on polyphenol-EGs as animal fat replacers using FT-IR with an ATR device.

5. Conclusions

The notion of improving the lipid component in meat products has been gaining attention in recent years giving rise to the development of new reformulation strategies to obtain healthy meat products and optimal technological properties. New procedures to create lipid materials based on structured lipids, such as EGs and OBAs, are especially interesting owing to their solid-like properties and the fact that they confer valuable nutritional and technological properties to meat products into which these materials have been added. EGs and OBAs have been employed successfully as animal fat replacers in the development of different meat products (sausages, hamburgers, etc.) with healthier lipid contents (qualitative and quantitative).

Application of vibrational spectroscopy (infrared (IR) and Raman spectroscopy) has significant possibilities and an increasing range of applications in the analysis of structured lipids (mainly EGs and OBAs) and meat products into which these lipid materials have been incorporated for the purpose of achieving healthier lipid contents. The use of vibrational spectroscopic techniques for both lipid materials and health-enhanced meat products have provided us with a greater insight into how the structure of their main components (proteins and lipids) impacts technological properties, mainly texture. This information can be helpful in choosing the most appropriate compounds for lipid materials used as animal fat replacers to obtain specific technological behaviors. Structural information and its correlation with certain technological properties could help in the selection of the most optimal lipid material and other compounds to obtain certain properties in health-enhanced meat products into which they are included.

Author Contributions: Conceptualization C.R.-C., and A.M.H.; data curation, A.M.H.; formal analysis; funding acquisition, C.R.-C. and A.M.H. investigation, C.R.-C., and A.M.H.; project administration, C.R.-C. and A.M.H.; resources, C.R.-C. and A.M.H.; writing—original draft, A.M.H.;

writing—review and editing, C.R.-C. and A.M.H. All authors have read and agreed to the published version of the manuscript.

Funding: This research was funded by the Spanish Ministry of Science and Innovation (PID2019-107542RB-C21), by the CSIC Intramural projects (grant numbers 201470E073 and 202070E242), CYTED (grant number reference 119RT0568; Healthy Meat network) and the EIT Food Project 20206 (https://www.eitfood.eu/).

Data Availability Statement: The data presented in this study are available on request from the authors.

Conflicts of Interest: The authors declare no conflict of interest.

References

1. Jiménez-Colmenero, F. Healthier lipid formulation approaches in meat-based functional foods. Technological options for replacement of meat fats by non-meat fats. *Trends Food Sci. Technol.* **2007**, *18*, 567–578. [CrossRef]
2. Jiménez-Colmenero, F.; Salcedo-Sandoval, L.; Bou, R.; Cofrades, S.; Herrero, A.M.; Ruiz-Capillas, C. Novel applications of oil structuring methods as a strategy to improve the fat content of meat products. *Trends Food Sci. Technol.* **2015**, *44*, 177–188. [CrossRef]
3. Herrero, A.M.; Ruiz-Capillas, C. Novel lipid materials based on gelling procedures as fat analogues in the development of healthier meat products. *Curr. Opin. Food Sci.* **2021**, *39*, 1–6. [CrossRef]
4. Pintado, T.; Ruiz-Capillas, C.; Jiménez-Colmenero, F.; Carmona, P.; Herrero, A.M. Oil-in-water emulsion gels stabilized with chia (*Salvia hispanica* L.) and cold gelling agents: Technological and infrared spectroscopic characterization. *Food Chem.* **2015**, *185*, 470–478. [CrossRef]
5. Herrero, A.M.; Carmona, P.; Jiménez-Colmenero, F.; Ruiz-Capillas, C. Polysaccharide gels as oil bulking agents: Technological and structural properties. *Food Hydrocoll.* **2014**, *36*, 374–381. [CrossRef]
6. Herrero, A.M.; Ruiz-Capillas, C.; Jiménez-Colmenero, F.; Carmona, P. Raman spectroscopic study of structural changes upon chilling storage of frankfurters containing olive oil bulking agents as fat replacers. *J. Agric. Food Chem.* **2014**, *62*, 5963–5971. [CrossRef] [PubMed]
7. Pintado, T.; Herrero, A.M.; Ruiz-Capillas, C.; Triki, M.; Carmona, P.; Jiménez-Colmenero, F. Effects of emulsion gels containing bioactive compounds on sensorial, technological and structural properties of frankfurters. *Food Sci. Technol. Int.* **2016**, *22*, 132–145. [CrossRef] [PubMed]
8. Poyato, C.; Ansorena, D.; Astiasarán, I. Linseed oil gelled emulsion: A successful fat replacer in dry fermented sausages. *Meat Sci.* **2016**, *121*, 107–113.
9. Pintado, T.; Herrero, A.M.; Jiménez-Colmenero, F.; Pasqualin Cavalheiro, C.; Ruiz-Capillas, C. Chia and oat emulsion gels as new animal fat replacers and healthy bioactive sources in fresh sausage formulation. *Meat Sci.* **2018**, *135*, 6–13. [CrossRef] [PubMed]
10. Ansorena, D.; Calvo, M.I.; Cavero, R.Y.; Astiasarán, I. Influence of a gel emulsion containing microalgal oil and a blackthorn (*Prunus spinosa* L.) branch extract on the antioxidant capacity and acceptability of reduced-fat beef patties. *Meat Sci.* **2019**, *148*, 219–222.
11. Domínguez, R.; Munekata, P.E.S.; Pateiro, M.; López-Fernández, O.; Lorenzo, J.M. Immobilization of oils using hydrogels as strategy to replace animal fats and improve the healthiness of meat products. *Curr. Opin. Food Sci.* **2021**, *37*, 135–144. [CrossRef]
12. Lucas-González, R.; Roldán-Verdu, A.; Sayas-Barberá, E.; Fernández-López, J.; Pérez-Álvarez, J.A.; Viuda-Martos, M. Assessment of emulsion gels formulated with chestnut (*Castanea sativa* M.) flour and chia (*Salvia hispanica* L.) oil as partial fat replacers in pork burger formulation. *J. Sci. Food Agric.* **2020**, *100*, 1265–1273. [CrossRef] [PubMed]
13. Herrero, A.M. Raman spectroscopy a promising technique for quality assessment of meat and fish: A review. *Food Chem.* **2008**, *107*, 1642–1651. [CrossRef]
14. WHO. *Diet, Nutrition and the Prevention of Chronic Diseases*; WHO Technical Report Series No. 916; WHO: Geneva, Switzerland, 2003; pp. 1–149.
15. Paglarini, C.S.; Vidal, V.A.S.; Martini, S.; Cunha, R.L.; Rosiane, L.; Pollonio, M.A.R. Protein-based hydrogelled emulsions and their application as fat replacers in meat products. A review. *Crit. Rev. Food Sci.* **2020**, 1–16. [CrossRef]
16. Zong, G.; Li, Y.; Wanders, A.J.; Alssema, M.; Zock, P.L.; Willett, W.C.; Hu, F.B.; Sun, Q. Intake of individual saturated fatty acids and risk of coronary heart disease in US men and women: Two prospective longitudinal cohort studies. *BMJ* **2016**, *355*, i5796. [CrossRef] [PubMed]
17. Wang, D.D.; Li, Y.; Chiuve, S.E.; Stampfer, M.J.; Rimm, E.B.; Willett, W.C.; Hu, F.B. Association of specific dietary fats with total and cause-specific mortality. *JAMA Intern Med.* **2016**, *176*, 1134–1145. [CrossRef]
18. Sacks, F.M.; Lichtenstein, A.H.; Wu, J.H.Y.; Appel, L.J.; Creager, M.A.; Kris-Etherton, P.M.; Miller, M.; Rimm, E.B.; Rudel, L.L.; Robinson, J.G.; et al. American heart association. dietary fats and cardiovascular disease: A presidential advisory from the american heart association. *Circulation* **2017**, *136*, e1–e23. [CrossRef]
19. McAfee, A.J.; McSorley, E.M.; Cuskelly, G.J.; Moss, B.W.; Wallace, J.M.W.; Bonham, M.P.; Fearon, A.M. Red meat consumption: An overview of the risks and benefits. *Meat Sci.* **2010**, *84*, 1–13. [CrossRef]

20. Siri-Tarino, P.W.; Sun, Q.; Hu, F.B.; Krauss, R.M. Saturated fatty acids and risk of coronary heart disease: Modulation by replacement nutrients. *Curr. Atheroscler. Rep.* **2010**, *12*, 384–390. [CrossRef]
21. Liu, A.G.; Ford, N.A.; Hu, F.B.; Zelman, K.M.; Mozaffarian, D.; Kris-Etherton, P.M.K. A healthy approach to dietary fats: Understanding the science and taking action to reduce consumer confusion. *Nutr. J.* **2017**, *16*, 53. [CrossRef]
22. FAO/INFOODS. *FAO/INFOODS Guidelines for Checking Food Composition Data Prior to the Publication of a User Table/Database-Version 1.0*; FAO: Rome, Italy, 2012; Available online: http://www.fao.org/fileadmin/templates/food_composition/documents/pdf/Guidelines_data_checking2012.pdf (accessed on 21 December 2020).
23. Dickinson, E. Emulsion gels: The structuring of soft solids with protein stabilized oil droplets. *Food Hydrocoll.* **2012**, *28*, 224–241. [CrossRef]
24. Jiménez-Colmenero, F.; Triki, M.; Herrero, A.M.; Rodríguez-Salas, L.; Ruiz-Capillas, C. Healthy oil combination stabilized in a konjac matrix as pork fat replacement in low-fat, PUFA-enriched, dry fermented sausages. *LWT Food Sci. Technol.* **2013**, *51*, 158–163. [CrossRef]
25. Heck, R.; Saldaña, E.; Lorenzo, J.; Correa, L.; Fagundes, M.; Cichoski, A.; de Menezes, C.; Wagner, R.; Campagnol, P. Hydrogelled emulsion from chia and linseed oils: A promising strategy to produce low-fat burgers with a healthier lipid profile. *Meat Sci.* **2019**, *156*, 174–182. [CrossRef] [PubMed]
26. Schrader, B. Infrared and Raman Spectroscopy. In *Methods and Applications*; Weinheim, V.C.H., Ed.; Verlagsgesellschaft mbH: Weinheim, Germany, 1995.
27. Herrero, A.M. Raman spectroscopy for monitoring protein structure in muscle food systems. *Crit. Rev. Food Sci. Nutr.* **2008**, *48*, 512–523. [CrossRef]
28. Damez, J.L.; Clerjon, S. Meat quality assessment using biophysical methods related to meat structure. *Meat Sci.* **2008**, *80*, 132–149. [CrossRef]
29. Herrero, A.M.; Carmona, P.; Jiménez-Colmenero, F.; Ruiz-Capillas, C. Applications of Vibrational Spectroscopy to Study Protein Structural Changes in Muscle and Meat Batter Systems. In *Applications of Vibrational Spectroscopy to Food Science*; Chalmers, J., Griffiths, P., Li-Chan, E., Eds.; John Wiley & Sons: Hoboken, NJ, USA, 2010; pp. 315–328.
30. Barth, A. Infrared spectroscopy of proteins. *Biochim. Biophys. Acta. (BBA)* **2007**, *1767*, 1073–1101. [CrossRef] [PubMed]
31. Guillén, M.D.; Cabo, N. Characterization of edible oils and lard by Fourier transform infrared spectroscopy. Relationships between composition and frequency of concrete bands in the fingerprint region. *J. Am. Oil Chem. Soc.* **1997**, *74*, 1281–1286. [CrossRef]
32. Guillén, M.D.; Cabo, N. Infrared spectroscopy in the study of edible oils and fats. *J. Sci. Food Agric.* **1997**, *75*, 1–11. [CrossRef]
33. van de Voort, F.R.; Sedman, J.; Russin, T. Lipids analysis by vibrational spectroscopy. *Eur. J. Lipid Sci. Technol.* **2001**, *103*, 815–840. [CrossRef]
34. Fraile, M.V.; Patrón-Gallardo, B.; López-Rodríguez, G.; Carmona, P. FT-IR study of multilamellar lipid dispersions containing cholesteryl linoleate and dipalmitoylphosphatidylcholine. *Chem. Phys. Lipids* **1999**, *97*, 119–128. [CrossRef]
35. Tuma, R. Raman spectroscopy of proteins: From peptides to large assemblies. *J. Raman Spectrosc.* **2005**, *36*, 307–319. [CrossRef]
36. Carmona, P.; Ramos, J.M.; de Cózar, M.; Monreal, J. Conformational features of lipids and proteins in myelin membranes using Raman and infrared spectroscopy. *J. Raman Spectrosc.* **1987**, *18*, 473–476. [CrossRef]
37. Levin, I.W.; Lewis, E.N. Fourier transform Raman spectroscopy of biological materials. *Anal. Chem.* **1990**, *62*, 1101A–1111A. [CrossRef] [PubMed]
38. Colaianni, S.E.M.; Nielsen, O.F. Low-frequency Raman spectroscopy. *J. Mol. Struct.* **1995**, *347*, 267–284. [CrossRef]
39. Maeda, Y.; Kitano, H. The structure of water in polymer systems as revealed by Raman spectroscopy. *Spectrochim. Acta Part A* **1995**, *51A*, 2433–2446. [CrossRef]
40. Beattie, J.R.; Bell, S.E.J.; Borgaard, C.; Fearon, A.; Moss, B.W. Prediction of adipose tissue composition using Raman spectroscopy: Average properties and individual fatty acids. *Lipids* **2006**, *41*, 287–294. [CrossRef]
41. Olsen, E.F.; Rukke, E.O.; Flatten, A.; Isaksson, T. Quantitative determination of saturated-, monounsaturated- and polyunsaturated fatty acids in pork adipose tissue with non-destructive Raman spectroscopy. *Meat Sci.* **2007**, *76*, 628–634. [CrossRef] [PubMed]
42. Ju-Yong, L.; Jin-Ho, P.; Hyoyoung, M.; Won-Bo, S.; Sang-Hyun, L.; Min-Gon, K. Quantitative analysis of lard in animal fat mixture using visible Raman spectroscopy. *Food Chem.* **2018**, *254*, 109–114.
43. Tao, F.F.; Ngadi, M. Recent advances in rapid and nondestructive determination of fat content and fatty acids composition of muscle foods. *Crit. Rev. Food Sci.* **2018**, *58*, 1565–1593. [CrossRef]
44. Nenadis, N.; Tsimidou, M.Z. Perspective of vibrational spectroscopy analytical methods in on-field/official control of olives and virgin olive oil. *Eur. J. Lipid Sci. Technol.* **2017**, *119*, 1600148. [CrossRef]
45. Yang, H.; Irudayaraj, J.; Paradkar, M.M. Discriminant analysis of edible oils and fats by FTIR, FT-NIR and FT-Raman spectroscopy. *Food Chem.* **2005**, *93*, 25–32. [CrossRef]
46. Fang, Y.; Dalgleish, D.G. The conformation of α-lactalbumin as a function of pH, heat treatment and adsorption at hydrophobic surfaces studied by FTIR. *Food Hydrocoll.* **1998**, *12*, 121–126. [CrossRef]
47. Lefèvre, T.; Subirade, M. Formation of intermolecular β-sheet structures: A phenomenon relevant to protein film structure at oil–water interfaces of emulsions. *J. Coll. Interface Sci.* **2003**, *263*, 59–67. [CrossRef]
48. Herrero, A.M.; Carmona, P.; Pintado, T.; Jiménez-Colmenero, F.; Ruíz-Capillas, C. Infrared spectroscopic analysis of structural features and interactions in olive oil-in-water emulsions stabilized with soy protein. *Food Res. Int.* **2011**, *44*, 360–366. [CrossRef]

49. Herrero, A.M.; Carmona, P.; Pintado, T.; Jiménez-Colmenero, F.; Ruíz-Capillas, C. Olive oil-in-water emulsions stabilized with caseinate: Elucidation of protein-lipid interactions by infrared spectroscopy. *Food Hydrocoll.* **2011**, *25*, 12–18. [CrossRef]
50. Muñoz-González, I.; Ruiz-Capillas, C.; Salvador, M.; Herrero, A.M. Emulsion gels as delivery systems for phenolic compounds: Nutritional, technological and structural properties. *Food Chem.* **2021**, *339*, 128049. [CrossRef] [PubMed]
51. Carmona, P.; Ruiz-Capillas, C.; Jiménez-Colmenero, F.; Pintado, T.; Herrero, A.M. Infrared study of structural characteristics of frankfurters formulated with olive oil-in-water emulsions stabilized with casein as pork backfat replacer. *J. Agric. Food Chem.* **2011**, *59*, 12998–13003. [CrossRef]
52. Herrero, A.M.; Ruiz-Capillas, C.; Pintado, T.; Carmona, P.; Jiménez-Colmenero, F. Infrared spectroscopy used to determine effects of chia and olive oil incorporation strategies on lipid structure of reduced-fat frankfurters. *Food Chem.* **2017**, *221*, 1333–1339. [CrossRef]
53. Pintado, T.; Muñoz-González, I.; Salvador, M.; Ruiz-Capillas, C.; Herrero, A.M. Phenolic compounds in emulsion gel-based delivery systems applied as animal fat replacers in frankfurters: Physico-chemical, structural and microbiological approach. *Food Chem.* **2021**, *340*, 128095. [CrossRef]

MDPI
St. Alban-Anlage 66
4052 Basel
Switzerland
Tel. +41 61 683 77 34
Fax +41 61 302 89 18
www.mdpi.com

Foods Editorial Office
E-mail: foods@mdpi.com
www.mdpi.com/journal/foods